Webvertising

Edited by SCN Education B.V.

HOTT Guide

Edited by SCN Education B.V.
This series of books cover special topics which are useful for a business audience. People who work or intend to work with Internet - at a management, marketing, sales, system integrating, technical or executive level - will benefit from the information provided in the series.

These books impart how new technologies and sales & marketing trends on Internet may be profitable for business. The practical knowhow presented in this series comes from authors (scientists, research firms and industry experts, a.o.) with countless years of experience in the Internet area.

The HOTT Guides will help you to:
- Enlarge your knowledge of the (im)possibilities of Internet and keep it up-to-date
- Use Internet as an effective Sales & Marketing tool by implementing new technologies as well as future-oriented strategies to improve your business results
- Facilitate decision making on a management level
- Reduce research costs and training time

These books are practical 'expert-to-manager' guides. Readers will see a quick 'return on investment'.

The Publishing department of SCN Education B.V. was founded in 1998 and has built a solid reputation with the production of the HOTT Guide series. Being part of an international IT-training corporation, the editors have easy access to the latest information on IT-developments and are kept well-informed by their colleagues. In their research activities for the HOTT Guide series they have established a broad network of IT-specialists (leading companies, researchers, etc) who have contributed to these books.

Books already in print:
ASP – Application Service Providing
Mobile Networking with WAP

Webvertising

The Ultimate Internet Advertising Guide

Edited by SCN Education B.V.

Die Deutsche Bibliothek - CIP-Cataloguing-in-Publication-Data
A catalogue record for this publication is available from Die Deutsche Bibliothek (http://www.ddb.de).

Trademarks
All products and service marks mentioned herein are trademarks of the respective owners mentioned in the articles and or on the website. The publishers cannot attest to the accuracy of the information provided. Use of a term in this book and/or website should not be regarded as affecting the validity of any trademark or service mark.

1st Edition 2000

All rights reserved
© 2000 Friedr. Vieweg & Sohn Verlagsgesellschaft mbH, Braunschweig/Wiesbaden, 2000

Vieweg is a company in the specialist publishing group BertelsmannSpringer.

No part of this publication may be reproduced, stored in a retrieval system or transmitted, mechanical, photocopying or otherwise without prior permission of the copyright holder.

Printing and binding: Lengericher Druckerei, Lengerich
Printed on acid-free paper
Printed in Germany

ISBN 3-528-03150-6

Preface

Time to get on board

Making serious money on the Internet means relentless promotion of your business. How can you make this happen? This HOTT Guide will give you hands on advise. We selected articles that will help you to grow your business. Some articles contain provoking thoughts of professionals. Others reveal strategies. Both can stimulate your creativity.

The Internet has leveled the field. New entrants without much capital can move very quickly. Marketing expert Jay Conrad Levinson has adapted a guerilla marketing approach to the Internet. In his book 'Guerilla Marketing online' he gives a variety of ways you can professionally promote your business for next to no money. Banner ad advertising is still one of the best ways corporate web sites promote themselves. There are many low-cost packages available, either priced on a per click-through basis or on a per impression model. Per click-through buys guarantee you will get a set number of people to click on your banner and go to your web site.

But there is more. Lots more. E-zine ads and e-mail newsletters are the Internet's most affordable advertising options. But why not start your own e-mail newsletter? Nothing creates sales like your own Internet community. You will be collecting subscribers. This e-mail list is one of the most powerful marketing assets you have. You can send them announcements of new products, lower prices or new affiliations with other firms.

There are many examples of people who made it, by positioning themselves as an expert on the Internet. One of the pioneers in the Internet Telephony, Jeff Pulver, founded in 1995 Pulver.com. This was mainly a web site. But he extended his business with the founding of 'The Minute Exchange', a spot market for the telecom industry.

Preface

'There is no doubt', dr. Kevin Nunley wrote, 'the Internet really is the biggest gold rush of our lifetime. It is unlikely you or I will get another chance as big as this one to earn huge profits anytime in the next 100 years. Someday people will look back and judge us as one of two groups: those who didn't recognize the Internet revolution and missed the greatest chance of our age, and those who smartly made a place for themselves in the new business model that will dominate the future. It is time to get on board.'

Ing. Adrian Mulder
Content Editor

Adrian Mulder is an Internet journalist who writes for major business computing magazines. He combines a technical background with a vast experience in the computer and business trade magazine industry.

Acknowledgements

Many people and professionals have contributed directly or indirectly to this book. To name them all would be practically impossible, as there are many. Nevertheless the editors would like to mention a few of those who have made the production of this book possible.

Executive Editor for SCN Education B.V.: *Robert Pieter Schotema*
Publishing Manager: *drs. Marieke Kok*
Marketing Coordinator: *Martijn Robert Broersma*
Content Editor: *ing. Adrian Mulder*
Editorial Support: *Dennis Gaasbeek, Rob Guijt, Richard van Winssen*
Interior Design: *Paulien van Hemmen, Bach.*

Also, we would especially like to thank dr. Roland van Stigt for laying a solid foundation for the HOTT Guide series to grow on.

Contents

11	**Chapter 1: An Introduction to Online Advertising**
13	Why Internet Advertising?
	By Tom Hyland
19	Internet Advertising
	By Coshe.com
27	It's an Ad, Ad World
	By Lee Weiner
31	Webvertising in an Accidental Industry
	By Nocturnecom
35	Advantages and Disadvantages of Advertising on the Web
	The H.W. Grady College of Journalism and Mass Communication
39	BJ Webvertising Marketing Guide
	By BJ Webvertising
51	**Chapter 2: Online vs Traditional Marketing**
53	Channel One Banner Advertising Report
	By Channel One
75	Executive Summary: The State of One to One Online
	By Peppers and Rogers Group
79	Web-based Sales: Defining the Cognitive Buyer
	By Paul Zellweger
91	Interactive Relationship Marketing
	By David M. Raab

Contents

97	**Chapter 3: Optimizing an Online Campaign**
99	How Internet Advertising Works
	By Rex Briggs, Vice President, Millward Brown International, San Francisco, USA and Horst Stipp, Director, Social and Development Research, NBC Television Network, New York, USA
129	It Pays To Advertise. Effects of Business-to-Business Advertising on Decision-Makers: Results of Recent Research
	By American Business Press
139	The Seven Steps to Successful Direct Marketing
	By Carey Hedges, HN Marketing Ltd
145	Exploding the WEB CPM Myth
	By Rick Boyce
151	What Advertising Works?
	By Bill Doyle, Mary A. Modahl, Ben Abbott, Forrester Research
157	Increasing Advertising Effectiveness on the Web
	By Intel Corporation
167	Justifying the Web for Your Business
	By USWeb Corporation
183	Collection of DrNunley.com Marketing Articles
	By Dr. Kevin Nunley
211	**Chapter 4: Ways to Measure**
213	Counting "Hits" Not Best Measure Of Web Success
	By Steven Bonisteel
217	The Dirty Truth About Click Throughs
	By eMarketer "the authority on business online"

Contents

219	On-Line Advertising Campaign Measurement: How Cached Impressions and Varying Ad Serving Technologies Affect Reporting and Performance *By Nicole Goldstein*
231	Banner advertising more effective than tv or radio in luring web shoppers, according to Andersen Consulting survey *By Andersen Consulting*
233	**Chapter 5: Tips**
235	9 Ways to Write Sure-Selling Ads *By Binnie Perper*
239	10 Tips to More Effective Banners *By Nick Bullimore*
243	**Chapter 6: Webvertising In-Depth**
245	Designing Catchy, Effective Banner Ads *By Meredith Little*
253	The Bigger Picture - Free Vs. Paid Advertising *By Internet Marketing Company*
257	Technical White Paper: Advertising on the Web *By Tom Shields*
265	Generate and Tracking Response to Promotional E-mail *By Michelle Feit*
269	Why E-Mail Lists Have Come of Age *By Michelle Feit*

Chapter 1: An Introduction to Online Advertising

Why Internet Advertising?

Title: Why Internet Advertising?
Author: Tom Hyland
Abstract: Does Internet advertising belong on your media plan? This is the question everyone is asking. The answer most certainly is yes, regardless of the brand you manage or the catagorie in which that brand competes. These are the facts: television audiences are migrating to the Internet and the Net is the fastest growing medium in history.

Copyright: Internet Advertising Bureau (IAB)

Ironic, isn't it, that just 40 years ago television was considered "new media"? And just 15 years ago cable wore the same badge. During their respective early days, each of these "new media" had to prove their value to earn a spot on the media plan-the same position the Internet finds itself in today.

Does Internet advertising belong on your media plan? This is the question everyone is asking. CEO's are asking their brand managers. Brand managers are asking their agency account managers and account managers are asking their media departments.The answer most certainly is YES - regardless of the brand you manage or the category in which that brand competes. Look at the facts:

Fact:Television Audiences are Migrating to the Net

The erosion of the network television audience during the 1980s and 1990s changed media plans forever. In the early '80s, television was simple to plan and buy with just three networks to consider. Then came cable, then a fourth network called FOX, followed by a dizzying array of syndicated offerings and yet more new network entries: Paramount and the WB. New choices continued to fragment traditional television viewing and advertising budgets soon followed this trend.

Figure 1

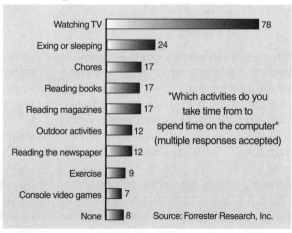

Source: Forrester Research, Inc.

Why Internet Advertising?

Television's recent history has demonstrated that media budgets ultimately are pragmatic. As audiences migrate, media plans follow, acknowledging that the ultimate goal of any brand is to reach its target audience effectively and efficiently. The exploding media landscape of the 90s-driven by increased TV audience fragmentation and the Web's popularity-have put this process into overdrive. Like the 80s and early 90s, media planners are, again, adapting their plans to account for the ever-growing numbers of people spending increasing amounts of time online at the expense of other media.

The first evidence of this audience migration appeared last summer in a Forrester Research report. The researchers asked PC users which activities they were giving up to spend more time on their computers. And, while 24% did admit giving up eating or sleeping to pound away on the PC, the activity sacrificed by over three-quarters of the respondents was television. Shortly after the Forrester findings were published, a study from The Georgia Institute of Technology's Graphic, Visualization and Usability Center (GVU) was released. This study, conducted on the Internet, asked users about their television viewing habits and what impact the Net might have on them. Their findings indicated a distinct shift in media habits with almost 37% of respondents claiming that they "use the Web instead of watching TV on a daily basis." Earlier this year, MSNBC noted the fact that Nielsen's February ratings sweeps found one million fewer U.S. households watching prime time television versus the same period last year. Simultaneously, Nielsen and CommerceNet released their Internet study, reporting that the North American online audience had doubled in the past 18 months. Clearly the conclusions of these two studies are far from coincidence. Taken alone, this migration of the television viewing audience to the Internet is particularly striking. This data is made even more impressive by the fact that Internet users are remarkably upscale. So, not only are we witnessing a fundamental shift in media habits, the Internet audience represents that hard-to-reach, well-educated, high income population most coveted by marketers.

Fact: The Net is the Fastest Growing Medium in History

Internet advertising began in 1994, when the first banner ads were sold (Hotwired, October 1994) and the first commercially available Web browser, Netscape Navigator 1.0, was released (November 1994). In a recent study, Mary Meeker, Managing Director, Morgan Stanley, and her team of researchers closely examined the adoption rate of the Internet, contrasted to the three other major "new media" invented this century: radio, network television and cable TV. As a common metric, they examined the number of years it took or will take for each media to reach 50 million U.S. users. With television, cable and radio included for historical context, the growth of the Net is nothing short of remarkable. Meeker estimates the Internet will capture 50 million users in just five years. It took TV 13 years and radio 38 years to reach this milestone.

Why Internet Advertising?

Figure 2: Adawareness

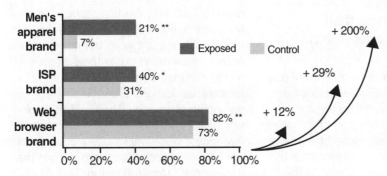

* Statistically significant at the 90% confidence level
** Statistically significant at the 95% confidence level

Fact: Internet Demographics are a Marketer's Dream

Every major research organization has studied the demographic composition of the Internet. While methodologies and approaches vary, the findings are consistent: Net users are young, well-educated and earn high incomes. And, increasingly, research shows that both men and women are using the Internet. Some topline findings from some of the more recent surveys are summarized here.

GENDER The March 1997 CommerceNet/Nielsen survey of Internet Demographics found that women now represent over 42% of the online population. Source: Nielsen/CommerceNet

AGE The average age of Web users is 34.9 years old, according to the 6th GVU WWW User Survey. This average age has been steadily increasing over the last several GVU surveys. (Fourth Survey: 32.7 years, Fifth Survey: 33.0 years, Sixth Survey: 34.9 years) Source: Georgia Institute of Technology, Graphics, Visualization & Usability Center (GVU), 1997

INCOME A 1996 survey by the Media Futures Program of SRI Consulting revealed that more than 65% of Internet users have household incomes of $50,000 or more, compared with 35% of the U.S. population as a whole (index 186). According to the sixth GVU study, average household income of Internet users is $60,800 (US). The distribution of income levels is very similar to the Fifth GVU survey: Less than $29K: 18.8%, $30-50K: 23.0%, over $50K: 41.1%. Sources: SRI International; GVU, 1997.

EDUCATION According to the same SRI study, more than 75% of Internet users have attended college, as opposed to 46% of the total U.S. population (index 163). Source: SRI International

Why Internet Advertising?

Fact: Web ad banners build brand awareness and may be better at generating awareness than television or print advertising.

Since their first appearance on commercial Web pages, the value of banner ads has been debated. Many felt they were physically too small to offer much branding and some advertisers convinced themselves that click-through was the only metric by which to measure ad effectiveness. They erroneously believed
- despite the fact that no research existed to support their belief
- that without a click-through, no brand building would occur.

In fall 1996, Millward Brown International set out to test the impact of banners on brand awareness, the first study of its kind. Millward Brown's objective was to measure the impact of a single ad banner exposure on brand awareness. The three brands tested included a men's apparel brand, a telecommunications brand and a technology company. The findings were significant and conclusive for each brand. Awareness was significantly greater among the banner-exposed (test) group than the non-exposed (control) group. Specifically, exposure to the ad banners alone increased brand awareness from 12% to 200% in a banner-exposed group. The study also compared the impact of the banner ads in this test to television and magazine norms

Figure 3: Adoption curves for various media - The Web is ramping fast

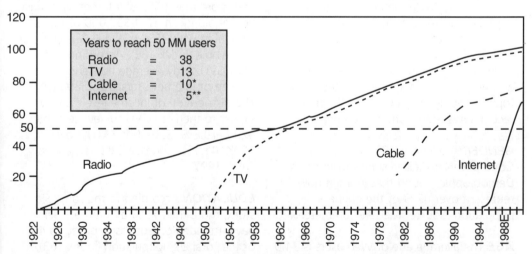

Source: Morgan Stanley Research, E=Morgan Stanley Research Estimate.
* The launch of HBO in 1976 was used to estimate the beginning of cable as an entertainment/advertising medium. Though cable technology was developed in the late 1940's, it's initial use was primarily for the improvement of reception in remote areas. It was not until HBO began to distribute its pay-TV movie service via satellite in 1976 that the medium became a distinct content and advertising alternate to broadcast television.
Morgan Stanley Technology Research Estimate

Why Internet Advertising?

from prior Millward Brown studies. The findings were remarkable: Single exposure to a Web banner generated greater awareness than a single exposure to a television or print ad. Millward Brown used their FORCE score ("First Opportunity to See Reaction Created by the Execution") as a means of evaluating the impact of the ad banners relative to other media.

A FORCE score indicates the effects of time, exposure weight, diminishing returns and base level. As such, FORCE scores can be directly compared across media types. As the median FORCE score for television advertisements is 10, the scores reported in the chart below (with an average score of 20 for the Web banners tested) suggest that Web banners tested very favorably to most TV ads, in terms of creating brand-linked awareness.

Now You Know the Facts

Every new medium has had to prove its value to advertisers. Just 15 short years ago, cable television fought to earn the respect of advertisers. Today it is a $6 billion industry. Those of us in Internet Publishing realize the Internet is no exception. We accept the challenge to prove the value of this medium and will build the case with facts-not hype- for including the Internet on your media plans. All these facts in aggregate create an undeniably compelling case for advertisers today to include the Internet in their media plans. As Lynn Upshaw, author of Building Brand Identity, noted recently, "The World Wide Web will be one of the strongest brand building tools available." Based on the facts at hand, we couldn't agree more.

Acknowledgments: The IAB would like to thank the following organizations for permitting us to reprint their research and quote from their analysis: Morgan Stanley, Forrester Research, MSNBC, The Georgia Institute of Technology's Graphic, Visualization and Usability Center, SRI International, CommerceNet/Nielsen and Millward Brown International.

Internet Advertising

Title: *Internet Advertising*
Author: *Coshe.com*
Abstract: *The potential of the Internet to provide an efficient channel for advertising and marketing efforts is overwhelming, and yet no one is really sure of how to best utilize the Internet for these purposes. It is clear that the costs, strategies and effectiveness of Internet marketing differ greatly from conventional marketing. As businesses use the Internet more, and the Internet users become more accustomed to marketing activities, Internet marketing is becoming more and more popular.*

Copyright: Coshe.com

The Internet is the name for a group of worldwide information resources. These resources are so vast as to be well beyond the comprehension of a single human being. The increasing popularity of the Internet as a business tool can be attributed to current size and projected growth, as well as its attractive demographics. The potential of the Internet to provide an efficient channel for advertising and marketing efforts is overwhelming, and yet no one is really sure of how to best utilize the Internet for these purposes. It is clear that the costs, strategies and effectiveness of Internet marketing differ greatly from conventional marketing. What is the Internet-- what we call the Internet today has its roots in a network set up by the United States Department of Defense in the early 1970's. This network was established by the Advanced Research Project Agency. ARPAnet was connected to various military and research sites and was itself a research project into how to build reliable networks. In particular, research about how to build networks that could withstand partial outages, such as bombings, and still function was important. The theory was to create a system that required a minimum of information from computer clients, and allowed dissimilar computer systems to communicate with one another. This eliminated the need for computer users to carry magnetic tapes and insert them in order to transfer data from one computer to another. Now the computer could put its data into an envelope and address it to send a message on the network. The philosophy was that every computer on the network could talk with any other computer. These methods were so successful, that many of today's networks adopted these standards. Although there has been much controversy surrounding the use of the Internet as an advertising medium, advertising agencies have not stopped talking about it until 1992, when the Web became user-friendly, Bob Metcalfe of Infoworld magazine said, APaid advertising, mostly as we've come to know and love it since the early settlement of Madison Avenue, can save the Internet. Although this view is a bit extreme,

Internet Advertising

Metcalfe has some validity to his point. Paid advertising may make the Internet sustainably inexpensive, like television, newspapers, and other traditional media. Advertising on the Internet may also provide practical incentives for attracting the attention of Internet citizens to the information they need. Metcalfe expands this argument by indicating that the Internet is growing rapidly (Metcalfe, 1995). If indeed advertising can save the Internet, as Metcalfe argues, where is Madison Avenue in all of this? Slowly, advertising agencies are becoming aware of the Internet as a marketing vehicle. Yet the $138 billion advertising industry seems unprepared for what lies ahead. The ad business is still recovering from the severe recessions of the late eighties, and layoffs and cutbacks have left advertising agencies without the resources to invest in interactive technologies. A few of the bigger agencies are formally and actively looking into interactive media. But most efforts are small scale and highly theoretical, according to a recent Wall Street Journal article (Smith and King 1995). Interestingly enough, coming late into the information superhighway could be especially dangerous for advertising agencies, which have traditionally held strong stances among the media. Martin Nisenholtz, senior vice president and director of electronic marketing at Ogilvy and Mather is quoted as saying that ad agencies Aare going to get killed if they don't wake up soon. It's the largest single case of denial I have ever seen (Smith and King). It seems that many of the agencies are signing up with online services such as prodigy and America Online, yet fewer are willing to establish an Internet presence.

Those that are on the Internet, such as Poppe Tyson, are starting to realize that more consumers are getting their information from the Internet. AWe want to put on our clients on it, says Walt Guarino, general manager of Poppe Tyson. AAnd if we're not in it, we look dumb (Goldman 1994). Commercial businesses are the fastest-growing segment of the Internet because it provides a space in which you can gather information, communicate, and actually transact business. As businesses use the Internet more, and the Internet users become more accustomed to marketing activities, Internet marketing is becoming more and more popular. Marketing on the Internet involves both research and an active outflow of information. Marketing research is common on the Internet, where attitudes are tested, conversations actively pursued, and opinions solicited from many groups. Marketing plans are increasingly counting on Internet access for success. One of the prime uses of the Internet is in the area of customer support. Customers can reach a company on their own schedules, day or night, and obtain information. The customer support information only has to be transferred to an archive once, and yet it may be accessed by thousands of customers and potential customers. This is indeed a labor-efficient and cost effective method of distributing information. In addition, a business with a presence on the Internet is perceived as modern, advanced, and sophisticated. In these days of a highly-competitive global marketplace, the company that can reach and satisfy customers will have an advantage; and the Internet can help in maintaining relationships with customers. The Internet is also a fast and efficient way of networking with vendors and suppliers. With its global reach, the Internet can assist businesses in locating new suppliers and keeping in better touch with them to aid, for example, zero inventory planning. A business might locate and coordinate with suppliers in Taiwan, Norway, and New York. And the Internet system in some countries is often more stale than telephone service, which is often less reliable and less convenient. Maintaining up-to-date postings of a company's product information and prices also allows vendors to have continuous access to information that is needed in order to promote and sell your products. Small suppliers find that they can compete with larger industries by being easily available via the Internet. In a business where the concept of getting closer to the customer is prevalent, the Internet is becoming increasingly important. Internet-assisted sales, where customers are sought and served on-line through Gophers and a variety of virtual storefronts, are also becoming more popular. Customers are sought before the sale and supported after the sale. Customer and product support and technical assistance by way of the Internet is time efficient. Many companies provide e-mail assistance, including both individual and automated replies to e-mail questions and requests for information. Technical sheets, specifications, and support are offered through Gophers and FTP. Relationships with vendors and outlets are maintained via the Internet. In some cases, companies are doing actual product sales on the Internet. In addition, if the product is amenable to Internet delivery, as with software and information, the actual product is delivered via the Internet. Some companies are arranging product delivery through the Internet, where companies can create and support actual distribution channels. In a recent study conducted on agencies' use of the Internet and online applications for marketing, ninety percent of the participants interviewed responded that their agencies were indeed currently researching online and Internet applications of advertising. Sixty percent of the agencies interviewed had not developed a separate division to handle the new technologies, and only forty percent had developed a new division. This would lead me to believe that the current roles of account management, research, media planning, and creative are burdened with the extra task and weight of discovering and managing the new technology, or that the clients are handling the new technology in-house. Sixty-seven percent of the agencies were aware of at least one of their clients who was using or had used an online application for advertising. Of those clients. Of those clients, the categories where their products/services fell were: transportation-31%, retail-24%, food and service-21%, technological/communication-10%, financial services-8%, entertainment-3%, and media-3%. An overwhelming number of participants indicated that the online advertising media of choice was Prodigy Services. However, many indicated America Online. Traditionally, America Online was not allowed for advertising, but they have allowed for corporate sponsorships of chat room events as well as strategic research. It seems from these results that these agencies are considering sponsorships and advertising communications effort. As with

Internet Advertising

Online advertising, seventy-three percent of agency participants were not aware of any of their clients using or have used the Internet for advertising. Twenty-seven percent of respondents were aware of their clients who have used the Internet for advertising, and these clients fall into three categories: technological/communication industries-44%, transportation-33%, and retail-22%. The majority of agency respondents were not aware of how their client used the Internet for advertising. Home pages, databases, and information/FTP sites were popular methods of marketing on the Internet. Newsgroups and bulletin board systems were also used, though not as extensively as the other methods. In terms of perceived effectiveness, technical products and communications vehicles were rated as effective while transportation and retail were rated as unclear to whether they were effective or not. Overall, Online applications for advertising are perceived to be very effective, while Internet advertising is dragging behind. Clients are indeed demanding that their agencies be innovative and find new ways to disseminate messages. This challenge was the first publicly articulated by Edwin L. Artzt, chairman of Procter and Gamble, when he told a convention of ad-agency executives that traditional mass-market ads could be endangered. Likewise, Gerald Levin, CEO of Time Warner, told the same conference of the American Association of Advertising Agencies in White Sulphur Springs, West Virginia that agencies should become more active in the interactive business. Levin noticed that while clients who are involved with interactive media are often enthusiastic, Aagencies don't always share their enthusiasm (McCormack 1994). In a survey of senior executives conducted by Northern Illinois University for Advertising Age, twenty-eight percent of CEO's said that their agencies were not prepared for the information superhighway and twenty-seven percent said that they were not sure if their agencies were prepared. Marketing managers echoed this sentiment, with twenty-eight percent saying that their agency is ill-prepared, and thirty-one percent unsure. Furthermore, an overwhelming forty-seven percent of CEO's do not think that they will need an advertising agency if the superhighway will permit them to talk directly to their consumers. Thirty percent of marketing managers agreed with their CEOs (Ward 1995). Every facet of conventional marketing strategy is challenged by the evolving communication potential of the Internet. This mandates that companies develop new, or at least revised methods of marketing. but not only is there no one who understands all of the Internet, there is no one who even understands most of the Internet. The growth of the Internet is overwhelming. Now that most four-year colleges are connected, people are trying to get primary and secondary schools connected, along with local libraries. People who have graduated from college and know what the Internet is good for are talking to their employers about getting corporations connected. The Internet is growing at a staggering rate of about ten percent a month as more colleges, universities, and businesses come online. Depending on which source you choose to use, the estimated number of users of the Internet varies between 5 and 35 million. In the early 1990's, the network

Internet Advertising

was opened up to a few large commercial sites, and the international Internet was expanding rapidly. The growth of the Internet can be attributed to several factors, not the least of which is the open availability and affordability of the medium. All of this activity points toward continued growth, networking problems to solve, evolving technologies, and job security for networkers. There are several reasons businesses would be interested in the Internet as a communication vehicle. The Internet eases communication, corporate logistical concerns, globalizes local businesses, helps to gain and maintain a competitive advantage, aids in cost containment, enhances and assists collaboration and development, promotes information availability, encourages marketing efforts, assists in the transmission of data, and establishes a corporate presence in the progressive interactive world (Ellsworth and Ellsworth 1994). The development of the World Wide Web mosaic browsers has furthered the expansion of Internet technology and use. The presentation of information on the Web is much more friendly than the older, more traditional methods, and the uniform interface provided by Web-browsers have facilitated a user-friendly environment (Smith and King 1995). This friendly atmosphere, combined with the ability to use any of the Internet's tools within the hypertext culture, has been a catalyst for businesses to rush to the Internet in record numbers. Although there is little evidence that anyone is getting rich yet from this method of marketing, companies are flocking to the Web in anticipation of future payoffs. Some companies build and maintain their own home pages, while others spend thousands of dollars a month to let experts like Compuserve create the cyberspace-equivalent of fancy front doors and reception areas. While the Internet is doubling in size every year, the Web is doubling in size every few months, making this medium even more attractive to marketers (Goldman 1994). The citizens of the Internet have strong opinions about how the Internet should be used. To develop and maintain a positive image and customer acceptance, businesses need to be aware of the acceptable uses and customs of the Internet. Since the broadcasting of advertising on virtually all networks, the only way marketing on the net will work is if the Internet community has a good impression of, and wants to patronize your service. AGiving back to the Internet is an important concept for prospective Internet marketers. The Internet custom is that you can market your good sand services if you return something of genuine value. A similar concept, and closer to marketers MBA backgrounds, is that of Avalue added services, again pointing to the idea of obligation to provide something to the network community (Metcalfe 1995). Creating a business presence on the Internet is a multi faceted endeavor. There is obviously much time and effort involved in developing a marketing presence on the Internet. Although it can involve almost every Internet tool, it is often more successful when it is based on a more modest approach. There are numerous ways to create a business presence on the Internet. How a business chooses to use the Internet is simply a function of many factors including planned business goals,

Internet Advertising

the marketing plan, and the level of market penetration desired. Using the Internet as an entrepreneurial tool is an audience participation sport; two-way communication is the valued and expected norm. The Internet encourages interaction, and encourages customers to be providers of information as well as users. One of the major challenges posed to Internet marketers is the method by which they communicate. In traditional media, the firm provided a message to the medium, which disseminates it to a mass market of consumers. In the new medium, the consumers and firm can interact with the medium, as in ANet surfing, as well as provide the medium, to the point of setting up their own Internet servers. The most radical departure from traditional marketing environments is that consumers can provide product related content to the medium. Marketers must reconstruct the function of advertising to reflect this Amany to many communication method (Hoffman and Novack 1995). Responsible communication on the net is characterized by two marketing strategies; the marketing activity should further the development of the medium itself, and the activity should work with consumers to develop a shared understanding of the consumer benefits of the Internet. Internet citizens will be slow to attempt guerilla marketing strategies, and so knowledgeable marketers are inclined to do things according to the culture, needs and wants of the consumer (Verity 1994). As viewed on the Internet, advertising is intrusive, whereas marketing can be active and can provide valuable information and services as part of an effort to provide products and services. Businesses must make a paradigm shift from something highly intrusive and image oriented to something highly content-oriented in order to be successful on the Internet (Toffler 1980). The potential of the Internet and the Web is not only important to advertising agencies, but also highly influentuel in the retail market as well. Although many concerns exist about the prospect of the Internet as a retail environment, there has been a great number of retailers to establish sites on the Web. The potential customers these sights will bring retailers will help raise sales either directly, through Internet shopping, or indirectly, through product and company advertisement. As of yet, there is no sure way of knowing just how many customers are calling or walking into retail stores because of Internet marketing, but at the low cost, the potential for reaching a targeted audience is worth it. The following graph shows the projected Internet retail sales growth through the year 2010 (Sandberg 1996). The figures are based on the current rate of growth of the Internet and World Wide Web, coupled with current estimations of retail revenue through both direct Internet sales, such as the Acyber-malls, and retail store traffic generation increase from Internet advertising.

In setting up a Web site, there are a few things a retailer or anyone else for that matter, should consider in order to maximize the site's potential. First, the site is useless, if noone visits it. The best way to generate traffic in your site is the use of search engines. by registering your site with various search engines, Internet Asurfers will be provided with the name of your site in response to their searches for topics related to your site. In addition, Web links, which can greatly increase traffic on your site, can be set up throughout the Web, in sites related to yours (Sandberg 1996). Once the site is created and traffic is generated, retailers can begin to take advantage of the opportunity. Direct marketing of products, customer service, and surveys can all help to increase retail sales as well as help define just what demographics a site is reaching. The important thing to remember is that the average Internet user does not want to pay for everything you provide. However, retailers can still increase business by providing information for free to Internet users, and then, upon evaluating the who, what, and why of the potential customers they are reaching, either the retailer can refine his site to reach a more targeted audience, or retailers can begin to cultivate these new customers by providing up to date information on services and products (Verity 1994).. In creating a Web site, retailers should keep in mind some of the important things to avoid. First, retailers know that you cannot bore people into buying your product. Your site should not make the customer wait while its pages are built, and it should get to the point quickly. Next, think nothing of pride when dealing with your Web page, your focus should be the needs and desires of the customer, not what you think they should know about the customer. Thirdly, retailers as well as anyone with a Web page must keep it up to date. No one is going to want to visit your site, if its information is outdated. In addition, customers will sense if you are Ajust in it for the money. This is a turn off, and it should be avoided at all costs by not sending annoying, unwanted e-mail, or failing to offer the customer anything worth his time. Just as important, is the clarity of your site. A confused customer will not long be a customer. A retailer should keep his site simple and easy to understand. Finally, if a retailer really wants to successfully market on the Internet, it is imperative that the retailer learn about the Internet. Most Web sites are done by people with no marketing experience. This new medium offers exciting possibilities to those who will apply themselves and learn the strengths and weaknesses of the Internet. By sticking to these simple guidelines, retailers and marketers can make the most of what the Internet has to offer (Wells 1994).

Research concerning online and Internet applications concentrates largely on the consumer's perspective, awareness, and attitudes towards interactive media. The consumer studies that relate directly to attitudes and behaviors on the Internet are of particular interest to both advertising practitioners and consumer researchers. Unfortunately, there are no formal research studies concerning advertising agencies and their use of online or Internet applications. Most American adults do not know what interactive media are or how they could improve their lifestyles. Regardless of the future research conducted, it is clear that the Internet will provide a viable and cost-effective method of conducting business. Although there are currently several issues concerning the effective and appropriate use of this medium, it will no doubt become a powerful and intelligent communications vehicle for the future marketing efforts of corporations and retailers. Clearly, advertising agencies are expected to perform, and retailers have already begun to take advantage of the opportunity the Internet has to offer. They must either catch the marketing information wave or be left to drown in this fast-growing medium. This paper makes a strong argument for Asurfing on the net with caution, yet an adventurous spirit. The agency or retailer with the most information and the skill by which to use it will catch the strongest crest and enjoy the longest ride.

BIBLIOGRAPHY:

Ellsworth, J.H. And M.V. Ellsworth. 1994. *The Internet Business Book*. New York: John Wiley and Sons, Inc.

Goldman, K. 1994. Ad Agencies Slowly Set Up Shop At New Addresses on the Internet. *The Wall Street Journal* (December 29).

Hoffman, D. And T. Novak. 1995. Wanted: Net Census. *Wired*. 2.11 (November): 93-94

McCormack, K. 1994. Levin to Agencies: Get Interactive, Now. *Adweek* (May 23). Metcalfe, B. 1995. Advertising can save the Internet from becoming a Utopia gone sour. *Infoworld* 16 (May 30): 48

Sandberg, Jared. 1996. Making the Sale. *The Wall Street Journal* (June 17): R6 Toffler, A. 1980. *The Third Wave*. New York. Morrow.

Verity, J. W. and R. D. Hof. 1995. The Internet: How Will It Change the Way You Do Business. *Business Week* (November 14): 80 - 86, 88.

Ward Fawcett, A. 1995. Agencies don't make the grade. *Advertising Age* (August 1):18

Wells, Melanie. 1995. Desperately Seeking the Superhighway. *Advertising Age* (August 22): 14 - 18

It's an Ad, Ad World

Title: It's an Ad, Ad World
Author: Lee Weiner
Abstract: Now that advertising on the Web has become commonplace, it's time to impose some order. A few tips on how to choose your weapon. How do advertisers ensure that their ads are seen by qualified readers? And what about managing where and when banners appear on a site? How does a Web site owner persuade advertisers to advertise on his site, and how does he get them to continue to advertise?

Copyright: © 1998 CIO Communications, Inc.

There's no doubt that the web is an increasingly attractive medium for advertisers, most of whom salivate at the Internet's potential to reach an increasingly targeted segment of customers. And for Web site owners, there is the potential to take in considerable revenue by selling advertising space on their sites. At the same time, Web advertising presents new challenges. How, for example, do advertisers ensure that their ads are seen by qualified readers? And what about managing where and when banners appear on a site? How does a Web site owner persuade advertisers to advertise on his site, and how does he get them to continue to advertise? The answers are becoming increasingly easy to find. The place to look is advertising management software.

Much like its counterparts in traditional media, online advertising is all about numbers. When advertisers place banners on a Web site, there are two numbers that can be used to tell them how effective their ad has been. The first is the number of "impressions," or times the ad appears on a browser. The other is the "click-through rate," or the number of visitors to the Web site who clicked on the ad and were linked to the advertiser's own site. Based on these numbers, the advertiser can decide whether changes should be made in the campaign.

Most Web sites report both metrics to advertisers using some type of advertising management software (or service). Also known as an ad server, this software is becoming increasingly important for both advertisers and site owners.
When advertising first appeared on the Web, site owners developed their own systems to calculate the metrics advertisers demanded. These were proprietary systems, and not all calculated the same numbers the same way. Things like caching, frames and other Web technologies affected the metrics the systems produced. The result, predictably, was confusion. It wasn't long before software companies caught on to the need for some standardization and to the

It's an Ad, Ad World

opportunity to develop systems that help manage the process.

Today, products such as DoubleClick or NetGravity serve many needs for both the advertiser and the site on which the ads are posted. First, they assure advertisers that impression and click-through metrics reported by a Web site are reliable. The online advertising world does not yet have standards; while these technologies do not necessarily create them, they do give a third-party audit of what is happening on a site. Second, ad servers assist in the management of scheduling, rotating and tracking banners. They also allow advertisers to target banners to specific users. How do you make sure that basketball fans see a banner for the ESPN SportsZone Basketball Section while webmasters see a banner from Microsoft touting its new Web technology? Most ad management software can enable sites with the technology to target ads to specific users. Advertisers are also buying key words, mostly on site search engines. Type "skiing" in Yahoo's (www.yahoo.com) search box, for example, and you may see a banner for a Colorado ski vacation. Finally, advertisers today are creating banners more technologically advanced than ever before, and ad servers are what allow Web site owners to accommodate and report on them. Whereas GIF 89a (or the animated GIF) has been an unwritten standard in the online advertising world, now advertising agencies are creating RealAudio (www.real.com), Shockwave (www.macromedia.com), Java and HTML banners. Banners now embed audio or some video and potentially some action

that can be initiated by the user (with a mouse). To be able to accept and track such banners, Web site owners need a system that can handle them.

More than a dozen ad servers are on the market today, so advertisers and Web site owners must make choices. For Web site owners, one critical question is whether to outsource. Anyone can, of course, purchase the software and install it on a Web server, but many vendors also offer services, or ad networks, that will "serve" your banners for you using their own servers and ad management software. Both options have pros and cons. Running the software in-house requires a staff that understands this stuff, but it offers more control and flexibility than outsourcing does. Going with an outside vendor avoids the headache of running the software, and your banners are served centrally and easily. The question to answer here is, Can your staff support the server software?

If you run your own software in-house, ads will be served by something called a server-side include. It is a tag that is put into a Web page that calls the ad server program to execute. It then puts a bit of HTML in the page dynamically. If you would like to be able to serve all kinds of banners from your Web site, including those using newer technologies such as Java and RealAudio, this is the way to go. Serving such banners using an ad network outsourcer can require placing them in frames or using technologies not supported by all browsers.

NetGravity (www.netgravity.com) of San Mateo, Calif., and Accipiter Inc.

It's an Ad, Ad World

(www.accipiter.com) in Raleigh, N.C., are two companies that develop ad server software. Of the two, NetGravity offers a few more features, but either system can handle large and small sites. In both cases, the administration piece is done completely via the Web, and the server can run on NT and Unix Web servers. NetGravity builds a great deal of flexibility into its system, including the ability to support such things as key word and demographic targeting. With the software, for example, you can serve ads to a certain demographic group based on the information that gets sent to a Web server when customers visit it. That information could be visitors' IP addresses, a number that can be traced back to their domain names, or it could be registration data that visitors give to the owner of the Web site. Using such information, the site can identify visitors who like to read, for example, and the ad server can show them banners from advertisers that sell books. NetGravity also offers extensive reporting options, allowing Web site owners to customize the reports they send to advertisers. Indeed, both products support reporting templates customized for certain advertising agencies that buy a lot of space on electronic media. Needless to say, this can save considerable time because reports will not have to be manually customized for each client.

Ad network, or outsourcing, services work somewhat differently. Web site owners implement an ad by uploading the physical graphic file (most commonly a GIF) to the server hosting the banner. Using the administration component provided by the service, you configure the ad to run for a specific number of impressions and a specific amount of time; in addition, you may configure it to run in a certain area of the site or to target it to certain users. To get the ad to appear on your pages you simply insert the correct HTML; this calls the ad server to fetch the banners and the correct click-through URL.

Most ad networks offer similar features. Virtually all can rotate banners, count impressions and click-throughs and serve enhanced banners, for example, as well as report on all activity on the site; in addition, many can target specific customers. The leaders in the ad network business are DoubleClick Inc. in New York City and AdKnowledge Inc. in Palo Alto, Calif., the result of the recent merger of Focalink Communications Inc. and ClickOver Inc. DoubleClick offers extensive targeting, allowing Web site owners to aim ads based upon such things as visitors' area codes or ZIP codes. The company owns a database of domain names linked with information such as where each domain is located, what type of organization owns it, how many people work there, etc. For example, if I work for Netscape Communications and I visit a site, the site's ad network will know that I am visiting from Netscape and that I am located in the Silicon Valley. Accordingly, it can show me banners targeted to people from the Silicon Valley area. Needless to say, this capability is of tremendous value to advertisers. Never before have they been able to put an ad in a place where it could potentially be seen by millions of viewers and say, "Show this only to people from New Jersey." In that way, they can get their message to the very people most likely to care.

While both DoubleClick and AdKnowledge offer similar features, they are administered in very different ways. DoubleClick uses a Web browser as the interface to manage the banners on your site. AdKnowledge's ClickWise, on the other hand, is administered via a client piece of software that is provided to you by AdKnowledge. Though DoubleClick's administration process is more strenuous, the fact that the Web is the interface is a big plus. It makes the service's client platform-independent and easily accessible from anywhere. This is online advertising-why not administer it using the same medium?

As advertising on the Web continues to grow, ad management systems are what will continue to allow everything to come together. In providing advertisers with the metrics, reports and features they demand, these systems are a key part of Web site owners' ability to keep advertisers buying space on their sites. Advertisers, of course, gain the ability to target as they've never targeted before. In addition, by tracking the number of impressions that are served where and when, ad servers can report how many ads a Web site can sell based on that historical data. The result is that Web site owners can better manage inventory and provide their sales teams with forecasts for how the site will perform. In short, ad servers let both advertisers and Web site owners do their jobs better, faster and cheaper-three good reasons to know who's serving what.

Web Developer Lee Weiner can be reached at lweiner@cio.com.

Webvertising in an Accidental Industry

Title: Webvertising in an Accidental Industry
Author: Nocturnecom
Abstract: The Internet, for all the miracles it has wrought in communications, is still a horrible way to buy a shirt. But it will get better, much better! In the world of advertising, ads on the Internet are a tiny drop in the bucket. In 1997, more than $185 billion was spent on all forms of advertising. That's $460 per capita, Just under $600 million was spent on Internet advertising, a mere $2 per capita.

Copyright: © Nocturnecom

Although the internet was never envisioned the have any commercial value (it was originally designed to to provide secure communications between nuclear silos accross the U.S.), it has become an important part of the mainstream corporate landscape. The internet still has a long way to go before it becomes a commercial success, but it is building speed towards that destination. Commercially, it has attracted the attention of companies with deep pockets that are willing to invest with venture capital. Many large organizations have discovered it as a means to reduce costs and improve customer and employee support systems.

The Internet, for all the miracles it has wrought in communications, is still a horrible way to buy a shirt. But it will get better, much better!

When I am using my internet browser and see the words "Java Loading" at the bottom of the screen, I cannot but fear an impending system crash This is one of the problems we face as technology in Beta becomes a daily part of our on-line lives. Baffling interfaces and limited selections make the corner grocery look like a miracle of organisation and choice comped to the most sophisticated on-line stores. For most consumers, today's Internet, far from being a perfect market, is the high street from hell.

In a world where consumers are faced with a bewildering choice of thousands of online stores, each one inadequate in its own way, just finding something to buy will be a triumph, never mind comparing prices and paying for it. Between the endless lists of online merchants and the delays as each graphics-heavy shopfront downloads, you can spend an hour just finding a product you are interested in. In the three years since the Internet has taken off, the slow growth of electronic commerce has been one of its greatest disappointments.

Currently most online stores are losing a fortune. Even MarketplaceMCI, the glossy online mall set up two years ago by MCI,

Webvertising in an Accidental Industry

an American telephone giant, has since shut its doors. The New York Times, which has a mature and well developed Internet publishing site, lost $15 million U.S. last year on it's online business. Knight Riddler, another major American Newspaper lost $16 million, while Tribute Co. newspaper publishers lost $30 million in 1997. Of course, these are the prices companies are willing to pay to establish themselves in this new medium.

Despite all this, future projections for internet commerce growth are positive. The wait for broad-based electronic commerce has been long, but it looks like 1998 may be the year that the market takes off. According to projections by International Data, a Massachusetts consultancy, by 2000 46m consumers in America alone will be buying online, spending an average of $350 a year each.

Few companies are as yet making money online, but plenty are trying. It is likely only a matter of time before they succeed. Andy Grove, the boss of Intel, the world's biggest chipmaker, recently summed up the online pioneers' attitude when asked about the return on investment from his firm's Internet ventures: "What's my ROI on e-commerce? Are you crazy? This is Columbus in the New World. What was his ROI?"

There are some success stories which pave the way for those brave enough to explore the commercial possibilities of this new medium. Place a store on the Internet anywhere and it is, in effect, everywhere. If you sells digital products—from software to information, it can delivered as easily as handing them over a counter. General Electric is saving a fortune by buying $1 billion-worth of goods from its suppliers online. Dell Computer is selling $1m-worth of PCs a day on the Web.

It is also possible to have an internet site that is not about gaining market share, but rather maintaining good customer relations. As customers, and clients come to expect on-line support for products and services, the demand for them will spur growth in the development of internet technologies catering to specific needs. For those who do not want to be on the internet, it may be a neccessity forced upon them.

In the world of advertising, ads on the internet are a tiny drop in the bucket. In 1997, more than $185 billion U.S. was spent on all forms of advertising. That's almost $460 per-capita. Just under $600 million was spent on Internet advertising, a mere $2 per capita.
In Canada, estimates of between 3 and 4 million dollars spent on Internet advertising would make this sector commercially insignificant if not for projected growth. The Canadian advertising industry beleives web-based marketing is in a high-growth phase and could hit $50 million annualy by 2000.
The ultimate promise of electronic commerce is undiminished, but the road from here to there is taking some surprising turns. Indeed, practically everything that was predicted about electronic commerce three years ago has turned out to be wrong.

Webvertising in an Accidental Industry

Figure 1: Bound to take off, forecasts of electronic-commerce turnover

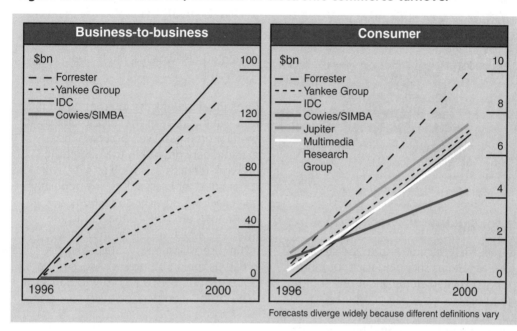

Forecasts diverge widely because different definitions vary

For starters, the big money is not in consumer shopping but in business-to-business commerce. This should not have been a surprise—it mirrors the physical world, where business transactions are worth about ten times as much as consumer sales—but few realised how quickly apparently stodgy firms would convert. The reason: most business transactions were already done at a distance, whether by fax, telephone, post, or private electronic links. Moving that process to the Internet makes it cheaper, faster and easier.

Only 3% of business-to-business Web sites are designed for direct sales, rather than for marketing and customer service, Even for consumer businesses, only 9% of sites offer online transactions.

A CommerceNet/Nielsen survey in March found that whereas 53% of Internet users in the United States and Canada had used the Internet to reach a decision on a purchase, just 15% carried out the final transaction on the Web. Yet it is just that last bit that is usually measured.

By the end of last year, 80% of America's Fortune 500 firms had a Web site, compared with only 34% a year earlier; but only 5% were conducting transactions on the Web, according to Forrester Research. Instead, their main reasons for setting up Web sites are to market their wares and help their customers, saving themselves money in the process.

Webvertising in an Accidental Industry

Figure 2: Indispensable

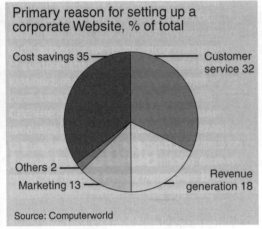

Primary reason for setting up a corporate Website, % of total

Cost savings 35
Customer service 32
Revenue generation 18
Marketing 13
Others 2

Source: Computerworld

In real life, where driving to the store can take as long as shopping itself, putting many unrelated stores under one roof is a good way to attract a critical mass of customers. Smaller stores, rather than rent space in one of a hundred obscure malls, are increasingly grouping themselves by theme, joining or creating consumer communities with shared interests. Some have taken to advertising in stores selling related goods, which is rarely done in the real world.

Consumers caught the Internet bug a mere four years ago, with the release of Mosaic, the first graphical Web browser. The extraordinary growth in the underlying network remains as strong as ever, with the number of users still nearly doubling each year. Worldwide, some 60m households are now connected to the Internet, which translates into around 120m users. Some estimates, taking a broad definition of Internet use, say that by 2000 the number could grow to 550m, or 10% of the world's population.

Figure 3: Here today, huge tomorrow

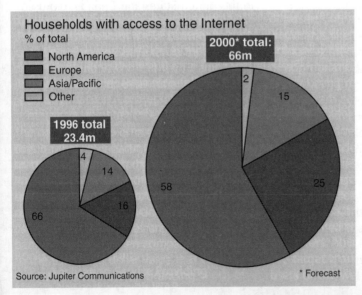

Households with access to the Internet
% of total

- North America
- Europe
- Asia/Pacific
- Other

1996 total: 23.4m — North America 66, Europe 16, Asia/Pacific 14, Other 4

2000* total: 66m — North America 58, Europe 25, Asia/Pacific 15, Other 2

Source: Jupiter Communications   *Forecast

Advantages and Disadvantages of Advertising on the Web

Title: *Advantages and Disadvantages of Advertising on the Web*
Author: *The H.W. Grady College of Journalism and Mass Communication*
Abstract: *Because of the ease of editing HTML files, changes to web pages or ads can be made at any time. No other advertising medium can compare. On a cost per thousand basis, web cost are very low compared to other media.*

Copyright: © H.W. Grady College of Journalism and Mass Communication

Objectives

- Know the advantages of advertising on the Web
- Know the disadvantages of advertising on the Web

As with all advertising media there are distinct advantages and disadvantages. These need to be understood in order to make an educated decision about where web advertising is viable for your product or service.

The Advantages

No lead time:
Because of the ease of editing HTML files, changes to web pages or ads can be made at any time. No other advertising medium can compare. Consider newspaper advertising which has at least a day's lead time, and magazines that have a lead time of one to two months.

User information readily available:
Many companies are now forming CGI scripts that allow the user to input data which is in turn accumulated in a database. This demographic and psychographic data can be very useful in determining a market for a product. However the main benefit is that you have a list of people that are already interested in your product. This is a great opportunity to send them a direct mail piece.

Creative Flexibility:
On the Web, it is possible to have sight, sound and motion. Using the technology available now, you can have sound clips, animated gifs and shockwave movies that all enhance the experience for the user. The possibilities are only limited by your creativity and the Web has the advantage of being extremely captivating. People will spend hours perusing a site that they find amusing and interesting.

Cost:
Implementing a home page can be very inexpensive. The costs involved with

Advantages and Disadvantages of Advertising on the Web

creating the home page includes labor costs and the cost of a server. While the initial outlay may be high, there are few maintenance costs. On a cost per thousand basis, Web costs are very low compared to the other major media. The cost of running an ad on another site is another story. Costs can fluctuate wildly, and if not done carefully, it is possible to waste a great deal of money, just as with all other major media. For example if you were advertising for a small store in a small town, it would not make sense to advertise on the Netscape page which can get millions of hits per day.

Selective Audience:
The Web is highly selective. Not only are the people using the Web very selective, the people using specific sites are extremely selective. It is possible to advertise anywhere that sells space to advertisers. This can be the homepage for a local grocery store, or auto dealer or a local newspaper. The Web has the ability to reach an extremely specific audience. Another benefit depending on your product or service is that people using the web tend to be college educated and have a household combined income of $80,000 or more.

The Disadvantages

Limited Audience:
The flip side of being highly selective, is that it is difficult to reach a large group of people. The people using the Web in general are still very specialized. Most people still don't even own computers, much less computers that have the

capabilities to access the Internet By advertising on the Web, you are excluding all of these people.

Difficult to determine effectiveness:
There is very little research available now in determining how effective a particular site is. As the Web grows, more and more marketing companies will become interested in these statistics. Also there is a difficulty with terminology. There is not a concensus on the best way to measure effectiveness. In particular, a hit versus a visit. The way the web is set up, it is possible to click on one homepage and it can count as ten or more hits to that site. So a visit is a more effective way to evaluate but more difficult to determine using current methods.

Technology not dependable:
As with all new technology, there are bugs that must be worked out. I'm sure you've been to several sites that crashed every time you tried to go there. Another disadvantage is that a lot of the new technology such as shockwave as well as others, require plug-ins that the user must download. Often instead of downloading it, they will just got to another page.

Clutter:
The Web is loaded down poor sites. It is extremely difficult for the typical user to find something of interest to them. Just as with any other advertising media, you must find a creative way to stand out among the rest.

Advantages and Disadvantages of Advertising on the Web

The Stop Button:
The more graphics used, the slower the page loads. If the page takes too long, the user hits the stop button. At that point you've lost all possibility of reaching that person. Make sure that your page is under 50K total so that the user can download it reasonably quickly.

Old browsers:
Because both Netscape Navigator and Microsoft Explorer are free to download from the web, it is extremely easy to have the most current browser. Even still, it is amazing how many people are still using Netscape versions 1.0 and 2.0. You can develop the coolest site with all of the newest technology, but if the users don't have the tools to access it, then your message is lost.

BJ Webvertising Marketing Guide

Title: BJ Webvertising Marketing Guide
Author: BJ Webvertising
Abstract: The following reports will help you to employ some of the Internet Marketing techniques to increase traffic to your website and how to make visitors return to your site again and again. The following subjects will de addressed: "Top 10 reasons for having your own website", "How to plan your new website?", "Guide to good web design", Choosing the right web hosting company", "How to choose your domain name wisely?" and "How to promote your website offline?".

Copyright: BJ Webvertising (http://www.bjweb.com)

"Top 10 reasons for having your own website"

The Internet provides a very powerful and flexible tool that can increase efficiency and productivity in addition to direct sales. The Internet can provide you substantial returns by reducing your costs while increasing capabilities.

1. Advertise More Effectively

We have all had the experience of seeing an advertisement on the weekend that perhaps piqued our interest. We intend to respond to the ad during "normal" business hours. Often, however, we never get around to it. Don't let this happen to your customers and potential customers. The best time to give them more information is RIGHT THEN, WHEN THEY WANT IT! The information they want and need will be on your Internet Web Site, 24 hours a day, 7 days a week. The customer may read your ad or see it on TV at 2:00 a.m. on a Sunday morning. If you display your Web Site address with your ad, they will be able to immediately call up your site and get the information they want, and have the ability to contact you or your sales staff. Your Internet Web Site address on all of your advertisements and promotions will do two things for you. It will give your company a cutting edge corporate image and it will encourage the viewer to take a look at what you have to offer. Since you are already paying for the advertisement, the addition of your web address increases your exposure without adding any cost.

2. Easily Provide World Wide Documentation At Little Cost

Distributing a traditional printed catalog can be very expensive. In addition to the high cost of production and printing, mailing costs can be enormous. Besides the fact that all too often mailed pieces can quickly become "junk mail". On the Internet, no matter how many people view your catalog, you will never run out. Printed catalogs quickly go out of date and have to be reprinted. Incurring all of the costs over and over and over again. On the Internet, when you have a change in your product line, offering, or pricing, you simply make a change to the site. Then instantly every customer has a brand new, up to date copy of your catalog. Perhaps best of all is that your entire product line is available worldwide without ever spending a cent on printing and distribution.

3. Give Your Company A Cutting Edge Corporate Image

Try this simple experiment the next time you are at a social function. Ask any business person or professional if their company is on the Internet? If it's not, they will likely be uncomfortable about the question. There is a sense of urgency about getting connected to the Internet. Most small business people act as though they feel guilty, and like they are the only one left not yet on the Internet. This pressure will only increase with time. Think back to just a few years ago when fax machines were thought of as amazing technology, but far too costly for most companies. Today a fax machine is not a luxury, but a business necessity. Your customers automatically assume you have a fax machine. Soon the same will be true of a Web site.

4. Make It Easy For Clients To Reach You

Your clients can instantly request more information. All they have to do is click one simple link or icon. This inquiry can be as simple as a request for more information or to have their name added to your mailing list. But it can easily be an extended survey of client opinions and feelings. An E-mail link puts you in instant contact with your client database. By capturing their E-mail addresses you can easily and automatically let your entire E-mail database of customers know about special offers or situations in seconds. Sending an E-mail broadcast to your targeted E-mail list can be done at virtually no cost.

5. Create An Effective Employee Information Site

Increase productivity and decrease branch office miscommunication. Your offices and employees anywhere in the world (and while on the road) can easily view up-to-the-minute changes in corporate policy or recent additions to your policies and procedures. This information can be secure, private information, or more general in nature. Every office can have up to date human resource policy information. Every office can be working on the same set of rules. This could prove invaluable as a guide to a manager in a new and potentially sticky situation. Since it's on-line, your entire policy manual could be searched within seconds. You would never face the problem of a remote manager or employee making a decision based on a policy that had been discarded long ago but is still "in the book".

6. Provide A Common Location For Emergency Communications

You can instantly communicate with your employees worldwide with last minute changes and instructions. This instant communication can be directed to your entire company, or to just a few select individuals. Most importantly, it can happen in seconds.

7. Support Different Platforms From Multiple Branch Offices

No matter what computer platform each branch office uses, you can easily communicate with all of them at the same time. The Unix-based data processing center on one region can easily communicate with the Apple-based graphics department on the other. Both of them can communicate with the Windows-based main office. The internet offers easy, inexpensive inter-office communication regardless of platform.

8. Long Distance Training Center

You can train or just update all of your remote branches and offices. The training can be received via weekly E-mailed sections. They can also be instructed to view the training section of your Web Site. The training can be done in conjunction with CD-ROM discs or just from the Internet.

9. Download Software And Updates

Dramatically reduce update times and avoid costly overnight express deliveries. As soon as updates are placed on the Internet, they are available for multiple users to download at the same time. The end user is happy because he or she can get the update within minutes of it being made available. You will be pleased when you see the savings on your overnight delivery service bill at the end of each month.

10. Remote Sales Force Tracking

Your sales force in the field can upload their client contact lists as fast as they can type them in. This can be done in the form of E-mail or you can have a template on your site. Each salesperson can fill out their information on line and with the click of a button insert it into the database. If security is important, passwords or encryption can be used to protect and authenticate information. You receive the information much more quickly and in a usable format. The sales representative can avoid unnecessary paper work by simply printing out their completed form prior to transmission.

"How to plan your new website?"

Why you must have a detailed plan even if someone else is designing your site.

There are many fine designers who are ready to build your new website. They offer a quick quote on the price of pages, good deals on graphics, maybe even search engine registration. But all these goodies don't do a thing to get you started. Before anyone can begin building your new Internet palace, you must have a detailed plan for what will be on your web site.

Starting a web site is a lot like getting your taxes done. The real work comes BEFORE you go to the preparer. Unless you've already got your box of organized receipts, the preparer can't fill in the forms. Start your web site plan by pinning down exactly what you want your site to accomplish. Draw a diagram on paper. List the pages you will have. Create some phrases to center copy around. You will save time, expense, and will get a much more effective site.

Will your pages sell your products and services online or serve as a detailed brochure to support your offline sales effort? Do you have one to two big products you will center your pages around or do you plan a big supermarket of products that need to be tied together under a prominent store image?

Focus Your Pages
Narrow, more focused web sites tend to do better on search engines. Their computers can easily figure what search terms to classify your site under. It pays to plan your site to be search engine friendly. Write down six to ten words and phrases that customers will use to search for you. Build a page around each of those. The title (that line in the box at the top or bottom of your browser), the meta tag (code in the page's html) and the text should mention your search phrase several times with copy that relates closely to the phrase. Most web site owners put up their new pages, THEN think about tailoring them for search engines. That means you end up redoing most of your pages with greater expense and poorer results. Your opening page should clearly tell visitors what your site is about. You may only have a few seconds to make your point before people click away. They should understand the most important benefit you provide. Use a headline and a related graphic to give people an instant image of what your site and organization can do for them. CarAccessories.com opens their first page with a graphic of car covers, mirrors, and fancy hubcaps along with the headline "Thousands of name brand accessories for your car...the largest selection on the Net...with fast online ordering." Readers instantly know what the site has to offer them.

Standard Pages That Build Customer Trust
Most sites have an "about our company" page and a "contact us" page. You can use these pages to build customer trust, one of the most important factors in getting sales online. Prospects trust you when they feel like they know you. Your "about us" page can feature a photo of you, your employees, your building, or anything else that gives a visual sense of who you are. You don't necessarily have to display a studio quality portrait. One man had a photo of his hand pitching a ball. Lots of successful home biz folks show themselves working behind a computer in a small, cluttered office. It is a scene their readers can identify with. Tell people why you do what you do, your company philosophy, and how you got started. Your "contact us" page should list the people in your company and provide several ways to contact them. Tell people why they should reach you and what they can expect when they do.

Support Your Main Theme With Secondary Pages
If you have a central product or service, introduce it with flair on your opening page. Save less important or secondary products for inner pages. If your site will have more than a dozen pages, gather similar pages into groups. Give each group its own gateway page that introduces the section and displays links to the related pages with a short description of what people will get when they click the link. One particulary organized client gave me a diagram of what pages he would have and how they would be grouped. Then he provided information on what each page should cover. He didn't write the copy, but

he did give me a solid idea of what he wanted on pages. It cut in half the time needed to build his big web site.

Simple, Clear Order Page
Keep your order page or pages as clear and simple as possible. You won't need a full-blown shopping cart if you offer five or fewer products. Make sure prices and descriptions are easy to understand. Anything that frustrates or confuses customers will make you lose sales. It is not unusual for a site to clean up its order pages and see an immediate surge in sales. Finally, resist the temptation to load down your pages with too many slow loading graphics. Keep your pages lean and mean. Slow loaders mean lost customers.

"Guide to Good Web Design"

What are the elements of a great website? We created a list - 12 rules for a great website. Use the rules as your guidelines when you design or overhaul your site.

Overall Look
Your visitor makes her assessment of your website within seconds of entering it for the first time. Go for clean, simple, and attractive. It looks professional that way.

Home Page
Your home page must convey your complete message. Think billboard. Tell me exactly what you want me to know right up front - tell me simply, clearly, and instantly.

Message
It is tough getting visitors. Give them a reason to stay. What is your offer? Exactly to whom is it directed? What makes your site better than others like it? What is in it for me if I stick around?

Speed
It is mandatory that your site load quickly. Use graphics judiciously. Get rid of the gimmicks. Try to keep each page and all its components under 40K bytes.

Graphics Size
Good graphics adds colour to your site. But graphics are your biggest bandwidth hog. So be careful. Crop them carefully, reduce their dimensions, and use compression software to squeeze out extra bytes.

Text Legibility
Don't get fancy with fonts. A simple font on a light background is usually best. Separate wide blocks of text into columns. Use plenty of white space.

Page Skimability
Use many short headings. Highlight keywords within the body of your copy to elaborate on the headings.

Copy Quality
Your web story is told with words - but your visitor is very impatient. He does not like to read. So make your copy simple, sharp and direct. Use half as many words as you think you should. Keep your paragraphs and sentences short. Use the word "you" a lot.

Navigation
Make it easy for your visitor to navigate your site - no matter where he finds himself. Include a link back to your homepage on every page of your site.

Contact Form
Your visitor does not like filling out your forms. If you must collect information from your visitors, collect as little as is necessary. If you plan on collecting your visitor's email address, tell her how you plan to use it.

Testimonials
Many of your visitors are suspicious of cheats and incompetent business people. Prominently list testimonials from your best customers to increase their confidence in you..

Mistakes
You will lose visitors if your site has broken links, missing graphics, or scripts that don't run correctly. Check your site often with the popular browsers (IE, Netscape, and AOL) and fix the mistakes.

"Choosing the Right Web Hosting Company"

Nearly 60 percent of Fortune 1000 companies surveyed said technical know-how is most important when choosing a Web hosting service. Forty percent look for reliability, and 34 percent look for the best price.

Price?
Cheaper is not necessarily better, but a large price tag doesn't mean your getting the best out there either. There are many start-up companies offering unbelievable prices. Be wary of "deals" that sound too good to be true.

Customer Services?
The Internet market is suitable for the old saying "you get what you pay for." A quality host should offer an online area with FAQs (frequently asked questions), guides, tips, and other resources. Can they help you find a designer or programmer for your site? Make sure you're getting your money's worth.

Technical Support?
This should be the number one consideration if you're not a technical guru. Is their tech support available seven days a week, 24 hours a day? How many members do they have on their support staff? How many customers do they have to support?

Services/Scripts/Software
A sharp hosting company should have a hefty library of scripts that you can use to add guestbooks, forms, statistics, counters and so on to your site. The host should also have support for Java, Shockwave, Cybercash, Real Audio, Real Video, VRML, secure transactions, and other utilities available to their customers.

Bandwidth?
A number of businesses have had to move their Web site because their host couldn't handle the number of hits, or charged exorbitant fees for hits above a certain level. What is the policy if your site becomes popular? How many other Web sites share your server?

BJ Webvertising Marketing Guide

Speedy Connections, Peering?
T3 lines, also known as DS-3, are a must. Anything less means the host, as a whole, runs slower; thus your customers will have to wait. Ask your host what the collision or saturation rate is. If it's over 50% it may cause problems.

Flexibility?
Does your host honor special requests or instructions? Can you start out with an economy package and then upgrade as your needs and budget increase?

Security?
What security features does your host offer or support? Many hosting companies claim to be secure, but when closely examined fall far short of their claim. Do they support "Adult Only" sites, SPAM, or other practices you might not want your company associated with?

"How to choose your Domain Name wisely?"

Domain Name can make or break your business! Sadly, most entrepreneurs today still do not appreciate the importance of their own Domain Name....

IT ALL STARTS WITH IMAGE
Your marketing strategy must also include branding your Web site, which is just as important as branding your company or product.

The reason for this is manifold. In today's world, we are constantly inundated with marketing messages. In his new book The New Positioning, Jack Trout states that a child in the UK will have seen over 140,000 TV commercials by the time he or she reaches 18 years of age -- and the US "is just warming up." The Internet is surely no different. It's literally filled with Web sites that range from sheer advertisements to others that are sponsored by them. Everywhere we turn, it seems, we are faced with some form of online promotional propaganda.

Our job as consumers has therefore become so immensely challenging that choosing a business from which to buy has become a dizzying process. For an online business to survive and thrive in today's hypercompetitive marketplace, it takes more than mere advertising to make a Web site successful (the kind of advertising that says "I'm open for business"). As marketing guru Dan Kennedy once said, "Institutional marketing is high-risk marketing," for the message needs to be repeatedly advertised in order to work -- if it ever does.

ELEMENTS OF A GOOD DOMAIN NAME
First, realize that a "good" domain name that sticks in the mind requires more than simply using a fictitious vanity name. However, it is imperative to note at this point that registered names have the ability to stick in the mind more effectively. Jack Trout once wrote that "The mind hates confusion, complexity, and change." Therefore, simplicity is of colossal importance since long or obscure URLs can be easily forgotten.

For example, rather than having a name with too many words, such as <http://www.domain.com/subdomain/your

name/~subfolder> or <http://names-with-too-many-hyphens.com>, you should get a very simple <http://www.yourname.com>. In fact, more and more companies and commercials are dropping the "www" from their URLs. Most Internet addresses can simply use "yourname.com," which is an even better alternative. In essence, the simpler it is, the better.

The importance of having your own domain name goes without explanation. It is the same as branding your business or product. But there are 3 reasons why you need a good, simple, and memorable domain name. First, there is the mnemonic factor. Instead of going through the inconvenience of numerous search engine results to get exactly what they want, most people will attempt to go to your site directly by guessing your domain name and typing a plausible URL in their browsers.

Mnemonics are words (or a combination of words) that are easy to remember. A repeatedly visited Web site is one whose URL, for example, includes the use of mnemonics. If it sticks in the mind, even if the URL is bookmarked, the site can be easily retrieved and will be visited often. "Yahoo!" <http://yahoo.com>, "HotBot" <http://hotbot.com>, and Time Magazine's "Time" http://time.com are perfect examples of mnemonics at work.

The second element is the credibility factor. People often associate long URLs with free Web sites or sites of lesser quality. People have a natural tendency to make what I call UPAs (or unconscious paralleled assumptions). In other words, if people notice that your site is hosted by a free or cheap provider, they will unconsciously assume that a parallel exists (i.e., that your product or service is just as cheap). Your domain name is like the headline of an article, and people will likely judge and visit your site according to its domain name.

Always remember that perceived truth is more powerful than truth itself. And a vanity domain name tends to heighten the perception of the Web site's value. As such, the UPA visitors will make with a domain name will often be one in which they conclude that the quality of the Web site will be as good as the name implies. Finally, the third reason is the the actual positioning process. If your domain name reflects your site's core benefit and instantly communicates how different you are from others, your URL will be positioned above the competition in the minds of your market. Since this element is the most important, let's deal with it a little further.

BENEFIT-BASED DOMAIN NAMES
People usually make a buying decision based on the kind of information that instantly communicates a specific benefit; one in which there is an implicit added value in making the purchase. Therefore, does your domain name intrinsically reflect the result or benefit of that which you provide and does so in an instant? It should. I am astounded to see many domain names that are still called by ordinary or blatantly unappealing names, such as with hard-to-spell words, numbers, abbreviations, or acronyms like "www.mgf.com."

Let's take the example of two different Web sites that promote similar products: Investments. One's address is "wealthwise.com" while the other "mgf-investments.com." Now, with all things being equal and when placed side-by-side, which site will be the one more likely to be chosen first? In essence, your domain name must be able to drive traffic to your site on its very own. It must also communicate how different and unique you are when compared to competitor sites, even before your site is ever visited.

As mentioned previously, people would far more want to skip the inconvenience of going through numerous search engine results. But if people do have to resort to an engine, their search will be greatly simplified and vastly more efficient if your domain name intrinsically reflects the core benefit if not the nature of your Web site. Remember that most searches are conducted by major topics or themes and not by names. Therefore, if your site's most popular keyword or benefit is within the domain name itself, that URL has greater chances of being in one of the top search engine results.

Therefore, play a word association game with your Web site. Look for the word or words that would instantly pop up in the minds of people when a need presents itself, a need your site likely fills. For example, http://freecoupons.com.my, <http://allergyrelief.com>, <http://morebusiness.com>, and <http://fastcar.com> are great benefit-based domain names that effectively create more top-of-mind awareness (and thus more traffic).

DOMAIN NAMES THAT DRIVE TRAFFIC
If the name you want is taken, then you can use your company or product's tagline (or part of it) as a domain name. A tagline is that small sentence that follows your business name, such as "You deserve a break today," "Roaches check in but they don't check out," and "It takes a licking but keeps on ticking." Great examples are <http://www.alwayscocacola.com> (a loyal Coca-Cola® fan site), <http://www.cavities.com> (Crest® toothpaste), and, of course, <http://start.com> from Microsoft®.

You can also use the site's main theme, feature, or product, even the site's nature or main business activity (i.e., what it does). Ultimately, choose a name that people can remember quickly and effectively so that, when you advertise among a thousand of your competitors, your URL stands out and sticks in the minds of the marketplace.

It is also a good practice to register variations of your name, including different spellings, product names, taglines, and associated words. One of the reasons for this is to ensure that these unused domain names don't end up falling into the hands of competitors. But more important, when people attempt to search for your site and enter a variation of your domain name they will still end up with your site as a result. It all boils down to the fact that your domain name is a fundamental marketing system in itself. Use it wisely and you'll see your traffic counter soar.

"How to promote your website OFFLINE?"

You just spent money, time, and research on developing a website. Now, you or your web designer or web company follow the standard rules of spreading it to the world-wide community –search engines, directories, banners, links, etc. After a few months, you've gotten some hits, but not quite up to your expectations. You consider some "tweaks" to your website, redo a few meta-tabs and do some checking on various search engines. You're there, but why aren't they coming?

There's A Real World Out There!
Most of us forget! We are inundated daily with information through newspaper ads, billboard ads, mail, phone, radio and TV. The total amount of time we are exposed to these forms of advertising is much greater than the time that we spend on the Internet. Advertising on the net, even in multiple locations is one-dimensional advertising. Other forms of advertising provide different formats for the same concept.

Start Simple!
Put the name of your website on your stationery, business cards, brochures, and invoices. List your website on any printed company related material no matter how benign it appears. Somebody might hang on to it and visit. One company had imprinted napkins with their logo and their website on it. Do you realize how often people write notes on napkins and take it along with them?

Send Announcements!
Many local papers accept press releases from all companies announcing new company developments. How about your new or redesigned website? On any type of announcement, list your website!

FAX The Facts!
Formal or informal fax cover sheets, and on all fax software cover sheets must have your website name.

Give Them A Reason To Come!
Design a contest, prize, or free tips, etc. for your site. Send out postcards inviting people to enter or use a special code to obtain their prize.

Use Phone Time!
Do you have a recording when customers are on hold? Why don't you tell them about your website and its features while they're on hold?

Add To Ads!
Any time you place an ad in a newspaper, journal, coupon, or even on the back of a register tape – put your website on it. If you're on the radio or TV commercials, tell them your website name. Your customers are everywhere and that's where you should be too!

Promote With A Premium!
Any type of advertising specialty, whether it's a pen, magnet, clock, cap, T-shirt, etc., needs your website on it. Any giveaway, whether used by an employee, salesperson, or by a customer becomes a walking billboard for your website. A simple pen that was printed " Stolen from

www.mycompanyname.com" was an incredible "hit"!

Add To Email!
Any time you send email include your website!

Keep Track!
Ask your customers where they heard about you and keep track of the most successful vehicle. It may surprise you!

Chapter 2: Online vs Traditional Marketing

Channel One Banner Advertising Report

Title: *Channel One Banner Advertising Report*
Author: *Channel One*
Abstract: *The Banner Advertsing Report is intended to provide on-line marketers with an in-depth understanding of the banner advertising market and the information needed to develop and execute profitable banner advertising strategies. The report is also intended to provide marketers with the knowledge to be able to deal confidently with interactive marketing agencies. It provides key figures on the banner- and on-line advertising industry, the market size and growth, banner and ad spending by industry, banner spending by model, banner advertising by advertiser, banner industry inventory, sites seeking on-line advertisers, click through rates and advertising rates,. The report studies the effect that banner ads have on branding. It compares banner to billboard advertising.*

Copyright: © Channel One ltd.

1. Executive Summary

Banner advertising provides marketers with the ability to boost brand awareness, enhance product perceptions, increase purchase intent and generate high volumes of traffic to the website quickly and cost effectively.

The Channel One Banner Advertising Report is intended to provide on-line markers with an in-depth understanding of the banner advertising market and the information needed to develop and execute profitable banner advertising strategies. The report is also intended to provide marketers with the knowledge to be able to deal confidently with interactive marketing agencies.

Section Two of the report provides key figures on the banner advertising industry, on-line advertising industry, market size and growth, banner ad spending by industry, banner spending by model, banner revenue by advertiser, banner industry inventory, sites seeking on-line advertisers, click through rates and advertising rates.

Section Three examines consumer attitudes and behavior to banner advertising.

Section Four studies the effects that banner ads have on branding. It compares and contrasts banner advertising to billboard advertising and examines the methodologies for creating brand linked impressions with banners.

Section Five of the banner advertising report analyses different banner advertising models- Cost per Thousand, Cost per Click-through, Cost per Transaction, Cost per Action, Cost per Lead and Cost per Sale.

Section Six focuses on the development of the banner advertising strategy concentrating on marketing strategy, market characteristics, sales, the target market, products, pricing, distribution and competition.

Section Seven covers budgeting for the banner advertising campaign with respect to different models.

Section Eight looks at targeting by demographics, psychographics, firmographics, content, activity, experience and technology. It also examines timing, frequency, and placement of banner ads.

Section Nine examines the different types of banners and standards. It outlines methodologies for developing banner messages and designing banners.

2. Banner Advertising Market

The banner advertising industry is an exponentially growing, fragmented market still very much in its infancy. It is characterized by huge variants in prices, unsold inventories, low penetration and lack of standards. Key information on the market is summarized below.

Figure 1: On-line advertising revenue breakdown

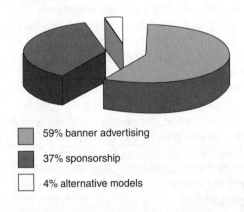

- 59% banner advertising
- 37% sponsorship
- 4% alternative models

source various

On-line Advertising Revenue Breakdown

Banner advertising accounts for 58 percent of the on-line advertising revenues, sponsorship for 37percent, interstitial ads for 3 percent and alternative models making up the remaining 2 percent according to the Internet Advertising Bureau (IAB). AdKnowledge puts banner ads at 60 percent, sponsorships at 37 percent and alternative models (including interstitial ads and advertorials) accounting for the final 3 percent. Simba reported 54 percent was spent on banner ads while 41 percent was spent on content sponsorship in 1997. Jupiter reported banner ads comprised 80 percent of all online ad spending in 1997.

Market Size and Growth

The On-line Advertising market has grown to a US $2 billion (IAB), US $1.5 billion (eStats) market in 1998 more than double the 1997 figure according to IAB. This accounts for 1.3 percent of the overall advertising market (InterMedia Advertising Solutions).

The revenue figure for 1997 at US $906.5 million (IAB), US $597.1million (Cowles/Simba), US $480 million (Jupiter Communications) was a sharp increase from the US $266.9 million (IAB) US $196 million (Cowles/Simba) and US $392 million (Jupiter) figure of 1996.

Internet advertising is expected to grow more than fivefold by 2002 according to Forrester. EStats predicts industry revenues will be US $3.8 billion by the year 2000 and US $8 billion by 2002. Veronis, Suhler & Associates predict US $6.5 billion by 2002. This will be only 3.7% of the total

Figure 2: Banner advertising growth

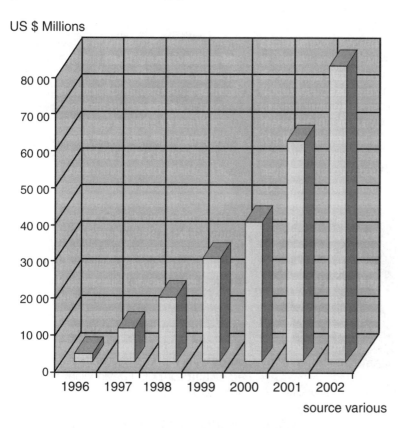

source various

advertising market which is expected to top US $215 billion.

Banner Advertising Spending by Industry
IAB found that computer related industries account for 26 percent of on-line ad revenue in 1998. Consumer related advertising accounted for 23 percent, financial services 13 percent, new media 13 percent and telecommunications 9 percent. InterMedia found that computer related industries accounted for 49.7 while financial services generated 8.5 percent of the revenues. They found that the medicines industry substantially increased their online advertising spending by 409 percent, and the local services and amusement sector by 223.9 percent. Microsoft spent the most on online advertising, at USD7.7 million, followed by IBM which spent USD7.6 million.

1997 figures from IAB show a marked change in 1998 where consumer related products accounted for 31 percent of total online spending, computing products 30 percent, financial services accounted for 18 percent, telecommunications 11 percent and new media accounted for 10 percent. Simba reported 32 percent of total money spent on advertising was consumer-related, 22 percent was on computer products, 20 percent on financial services, 7 percent on new media and 6 percent on telecommunications; national advertisers accounted for 88 percent of the total and the remainder was spent by local advertisers.

Figure 3: Banner advertising by industry

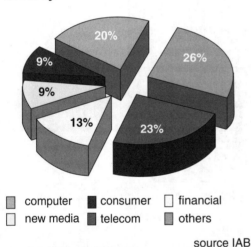

- computer
- consumer
- financial
- new media
- telecom
- others

source IAB

Banner Spending by Model
There are several models for the purchase/sale of banner advertising. These models are explored in detail further into the report. Hybrid (combination of cost per thousand and performance based) account for 56 percent of banner revenues, cost per thousand 40 percent and performance based 4 percent. (IAB)

Figure 4: Banner advertising by model

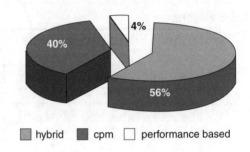

- hybrid
- cpm
- performance based

Banner Ad Revenue by Advertiser
The top two or three leading sites in individual content areas will get 80 to 90 percent of total advertising revenue according to Jupiter Communications.

Banner Advertising Inventory
80% of banner advertising inventory goes unsold. (Jupiter Communications). Adauction.com CEO David Wamsley says 50% to 70% of the ads on the top 500 Web sites go unsold. Fletcher Research reports only 14 percent of banner inventory in the UK is currently sold.

Sites seeking on-line advertisers
AdKnowledge found there was a 38 percent increase in the number of sites seeking online advertisers during 1998.

Click Through Rates
Click through rates have continued to decline. I/PRO reported in April 1996 that advertising response rates on Web varied between 2.4 percent and 17.9 percent. Netratings reports the decline from 1.5 percent in May 1998 to 1.1 percent in June

1998 to 0.9 percent in July 1998 to 0.7 percent December 1998.

Advertising Rates
AdKnowledge reports the CPM (the cost per thousand impressions) for May was $37.84, down from $40 a year earlier.

3. Consumer Attitudes and Behavior to Banner Advertising

Consumers are relatively accepting to banner advertising on the Internet. Research indicates that they realize that in most cases banners subsidize or fully pay for the cost of the websites that they visit.

INTECO found that 53 percent of Internet users in the US in 1988 believe that online advertising is necessary to keep the content of websites free. 43 percent said that they found banners less obtrusive than direct marketing and TV advertising. 23 percent did not find them less obtrusive. 22 percent said they would definitely pay more to subscribe to ad-free sites while 37 percent said they definitely would not.

MBinteractive found in their research for HotWired and Ogilvy & Mather that Web users support advertising on the Web. In their research for the Internet Advertising Bureau they went one step further and found that not only do web users tolerate advertising but some actually enjoy it!. When asked to rate their feelings to Web advertising on a 5 point scale ranging from "I hate it" to "its great", more than half their respondents to the survey reported a top-two box score.

The research by MBinteractive was conducted in 1997. At the time, the market was less flooded with banners and Internet users were less sophisticated in their browsing habits. More users clicked on banners as a way of finding new sites. Today many know where to find these sites and realize that clicking on a banner often brings them off on a tangent which adversely affects the time it takes to complete the task at hand. This may help explain the dramatic fall in click through rates over the last year.

Many Internet users now suffer from a condition called Attention Deficit Disorder. This is a term used to describe the decreasing attention spans and enhancement of filtering mechanisms by consumers who are exposed to higher volumes of input that they can process of act upon.

The average user is now exposed to over 100 banner advertisements during a 30 minute Internet session. They have less time to contemplate or process the banner advertisements they are subjected to. Advertisements that are not perceived to be of interest, relevance or value to the consumer are quickly forgotten. Forrester Research found that two thirds of a selected group of users could not remember the contents of the last banner ad they had seen.

Banner advertising is generally accepted and sometimes even liked if the banner provides the target market with access to the products and information that they are looking for. However, if the presence of banners adversely affect their browsing experience, for example, too many banners cluttered on a page resulting in slowing down the time it takes for the page to be viewed, Internet users will become dissatisfied with the publisher and maybe even the advertiser. Internet users suffer from Attention Deficit Disorder, to successfully execute a banner campaign the banners must highly targeted, relevant, memorable and outstanding for the target market.

4. Branding

Branding is a critical component of banner advertising especially because users don't always immediately click through to find out more information from the company's website. In fact only 0.7 percent of people who see a banner ad click on it on average, according to NetRatings. That's down from 2% in 1996, according IPRO and DoubleClick research.

To understand the power of banner branding on the information superhighway, banners can be compared to billboards along a regular highway.

- They are both constrained by the amount of information that they project to the user
- Both are driven or surfed by at high speed.

But if the right person sees the right image, the banner, like a billboard can create, change or instill a perception about the brand being advertised.

There is no doubt that a more detailed message relating to the product offering provides a greater amount of brand enhancement communication. Market researcher Ipsos-API Inc, found that Internet users were 46 to 63 percent more likely to remember larger, more complicated advertisements than the average banner ads. Interstitials ads which pop up in a new window were effective in conveying an advertiser's message 33 percent of the time, compared with 16 percent for banner ads.

However, when you look at the cost of banner advertising compared with Interstitial advertising this result is quite good. Banners advertisements not only have the ability to remind consumers about the brands for which they are already aware, they can inform users about new products or offers from the company. But more than simply boosting awareness, banners can actually impact the way consumers think about advertised brands.

Unlike billboards banner ad effectiveness seems to be driven by the fact that Internet browsing is an actively engaging exercise, similar to reading newspapers and magazines. Users are fairly attentive to the media environment - including the banner advertisements.

The effects of banner branding can be measured by asking focus groups questions about their awareness and perception of the brand. The best market research to date was conducted by MBinteractive for the Internet Advertising Bureau in 1997. The findings of the research are very interesting…

Channel One Banner Advertising Report

Advertisement awareness
After a single impression eleven out of the twelve banners tested show marked improvement in advertisement awareness. An additional impression of the banner boosted advertisement awareness by 30% on average (from 34.0% to 44.1%).

Shifts in brand perception
Five out of 12 banners demonstrate clear positive change in brand perception, while the sixth shows a polarization of positive and negative associations, with a positive net effect on purchase intent.

Value of the ad exposure
Among the twelve ad banners tested, on average, the value of the ad exposure is significantly greater than the value of the clickthrough. To put the clickthrough metric in proper context, consider that recall of the advertisement was boosted by four-tenths of one percent (from 43.7% to 44.1%) as a result of those who clicked on the ad banner. That's an increase of less than one percent! Said another way, 96% of the boost in Ad Awareness was caused by the ad exposure alone. The remaining four percent was caused by the clickthrough.

The findings of this market research conclude that click throughs should not be the only goal of a banner advertising campaign. Banner advertising can significantly increase the audience reach of a product, regardless of whether or not the consumer clicks on the advertisement, according to a recent study by NetRatings. It found that companies that have consistently advertised online have significantly increased their market share.

In one case, GoTo.com increased the Web banners advertising its site by 53 percent. The number of individual visitors to GoTo.com increased by 24 percent in the same time period. NetRatings believes the two figures are related.

Chasing click-throughs can in-fact be destructive to brand enhancement. Unbranded banners can boost clickthrough rates by appealing to the consumers curiosity. For example, banners with questions that urge the user to click to find out the answer can achieve significantly higher click-through rates than an advertisement that uses a branded message. However, the effects of the increased click-throughs will, in the majority of cases, not out weigh the effects that that a branded banner with lower click-throughs will have on the bottom line.

Creating Brand-Linked Impressions
Now that it has been determined that banner advertising should not be focused entirely on click throughs we can return to the comparison between banner and billboard advertising. It should be understood that more than 99 times in 100 the banner will be passed by, just like a billboard on a highway. Therefore the objective of the banner strategy should be the similar to the objectives of a billboard strategy, that is to…

Create a positive impression of the brand and lock it into the long term memory so the customer will draw on that memory when interaction with the brands category occurs.

Execution of the strategy will involve…

Attention
In the Internet environment where the consumer has developed filtering mechanisms to cope with information overload the first objective of the banner is to secure the consumers attention. Here the banner does not only have to compete with other advertisements for the consumers attention, they have to compete with the content of the webpages and the consumers desire to stay focused on finding the information that they are looking for.

Association
Once their attention is secured, the next step is to create and instill a positive association with the brand in the mind of the consumer. It is important to ensure that the positive association is linked tightly to the brand since it is possible for consumers to easily remember a particular banner but not be able to associate it with the brand. This type of banner although it helps the brands category, it does not help achieve the brand's objective. To achieve high levels of brand-linked association recall requires that the ad be memorable and what is memorable needs to linked to the brand.

Recall
Consumer interaction with the product category or brand should initiate the recall of a positive association with that brand, if the banner advertising campaign has been effective. Effective banner campaigns will ensure that the consumer is aware of the brand, will help the consumer associate the brand to the brands category, help him to remember the name or address of the website, to associate positive product perceptions with the brand when they see the listing in a search engine or directory and make an impact on the purchase intent of the consumer.

5. Banner Advertising Models

The banner advertising industry is still very much in its infancy. Standards and pricing models are now only emerging after heated debates on whether branding or direct response models should be used to calculate the pricing of banner advertisements.

The foundation of banner ad pricing (Cost Per Thousand) is based on the traditional broadcast model where pay per performance pricing is relatively impossible. With broadcasting, it is only possible to estimate the number of consumers that view the advertisement by using ratings which base their results on sampling. And since the medium only goes one way, consumers have to use a different channel to act upon the desire created by the advertisement. Therefore marketers can only estimate, not accurately measure the effects advertising on the target market.

The Internet brought interactivity to both consumers and marketers. Now marketers can not only accurately measure the number of consumers who see the advertisement, they can measure the response of the consumer to the advertisement from clicking on the banner, to interacting with the website, all the way to the sale. This functionality provided marketers with the ability to address the

Channel One Banner Advertising Report

return on investment of the campaign and brought with it direct response models to banner advertising (Cost Per Click, Cost Per Action, Cost Per Lead, Cost Per Sale.)

However, marketers are beginning to realize the benefits on banners on branding and see that click-throughs are just one measurable action of the target market to the advertisement. Banners can develop, change or lock a perception about the brand in the mind of the target market without ever clicking on the banner. Therefore the number of times that the banner is displayed (Cost per Thousand) is being used to estimate the effects of banner advertising on branding.

The effects of banner advertising on branding can, however, be measured using a direct response model. This would involve using cookies (putting a tag on the consumer every time they see an advertisement) to track the number of times they see the banner ad. When the consumer enters the website, and/or generates a transaction the advertisers could compensate the publisher(s) of the advertisement based on the number of times the consumer viewed the ad. This model has not been used on the Internet to our knowledge. It has been developed by Channel One to provide an alternative to the CPM model for branding.

However, there will still be opposition to direct response models from publishers on the grounds that they cannot control the creative, the offer or the product being advertised on their site. This will change, in our opinion as the market becomes more educated and publishers realize the substantial revenue that can be generated from tracking the advertisement to the sale.

In fact many of the large publishers are integrating sales based tactics into their strategies. Time Warner plans to sell CD's from Atlantic Records and videos from Warner Brothers. Disney is offering travel reservations to the visitors to it's theme parks. "If you can be a profitable business based on ad sales, and then have a successful e-commerce platform on top, then you will have a robust, powerful business going forward," says Scott Ehrlich, executive producer of News Corp.'s NewsAmerica Digital Publishing.

In conclusion smart advertisers and smart content owners now realize that it is important to choose the right advertising model in order to maximize return on investment. This will involve choosing a hybrid model that measures the effects of each variable within both performance based and CPM models. The winners will be those who choose the most profitable equation to determine the advertising strategy.

Break down of Banner Advertising Revenue by Model

56 percent of ads are sold as a package, success measured by a combination of impressions, click-through rates and cost-per-sale. 40 percent are sold as straight cost-per-thousand deals, and 4 percent as performance-based ads according to IAB.

Figure 5: Banner advertising by model

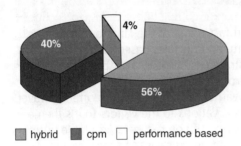

■ hybrid ■ cpm □ performance based

5.1 Cost Per Thousand (CPM)

The CPM (M being the Roman numeral for thousand) model allows the purchase and sale of banner advertising based on the amount of times the banner advertisement is displayed. Each time a user views a page that contains the banner ad counts as one impression. Banners are sold by the impression in packages of 1000.

5.1.1 Rates

The industry average CPM, according to software firm AdKnowledge, in September 1998 was US$36.29 CPM (which means it costs $36.29 to buy one thousand impressions). The figure is the lowest since they began tracking this data in December 1997 when it was $37.21 CPM. It has fallen $1.55 to $37.84 CPM since May '98. However, the market is still very fragmented and unsold inventory is auctioned off for as low as $3, while banners on highly targeted sites can be sold for up to $150.

CPM is currently the most popular model for banner advertising. It is used by content owners and banner networks who see their role as providing targeted audience for their advertisers. Many use this model because they do not want to get involved in the advertisers marketing objectives and argue that they do not have control over the creative (banner ad) the website (the pages the customer reads after clicking on the banner, the offer, the product or the company (the advertiser).

CPM is the best model in existence that can be used to measure the effects of branding. It allows the marketer to compare available media on three variables: the place (publishers site), the people (audience) and the price (CPM). The information gained from the comparison is used to forecast the effects of branding. It can also be used in conjunction with other models to forecast the return on investment.

5.2 Cost Per Click Through (CPC)

The CPC model allows the purchase/sale of banner advertisements based on the cost for each individual who enters the advertisers website to find out more information as the result of clicking on a banner.

CPC bought by click through networks such as ValueClick from media and content sites is $0.12 to $0.16 This means that they pay the media site 12 to 16 cents for each individual who clicks on the banner. This is an increase of $0.06 from last year. However the market is still very fragmented and CPC can range from 5 cents to $2 depending on the industry.

CPC is a pricing model developed in demand to the users need to know the return of investment before they execute a campaign. It's main flaw is that most banners are bought by CPM. Publishers can forecast the revenue generated from a campaign quite accurately dividing the number of banners by the industry average click through rate, which according to NetRatings is 0.7 percent. This means that 7 out of every thousand banners impressions are clicked upon. Therefore some clickthrough networks are purchasing banners at the equivalent of 7 X $0.12CPC = $0.84CPM.

With CPC it is important for publishers to be selective or become involved in the development of the creative, the message and the offer of the web site being advertised. This will ensure that the banners on the media site will sustain a high click through rate and therefore generate higher revenues.

However many CPC banners are purchased from unsold inventory drawing on the model that something is better than nothing. This model works pretty well since up to 80% of banner advertising goes unsold.

The bottom line, is that networks, publishers and advertisers will do the math and compare CPC with CPM in order to purchase or sell advertising at the most profitable prices. Most web advertisers don't prefer CPC pricing according to a survey by WebCMO in September 1998. The estimated preference for CPC -0.13 vs. 0 for CPM.

5.3 Cost Per Transaction (CPT)

CPT models allow the purchase/sale of banner advertisements based on the target market completing a specific objective after they click on the banner and are brought to the advertisers website. Models include Cost Per Action (CPA), Cost Per Lead (CPL) and Cost Per Sale (CPS).

With CPT it becomes more important for publishers to be selective on their advertisers and to integrate both advertising strategies to form a profitable and synergistic relationship. CPT banners are not purchased like CPM. Here the publisher does not market their available advertising space. Instead the advertiser must actively market their CPT offer to publishers and convince them that it is in their benefit to form a CPT advertising relationship.

CPT models require the advertiser to integrate software into their website that identifies which customer was referred to the website by which publisher. The customer is followed through the website until they make a transaction. The transaction is logged and a payment is credited to the publishers account.

5.3.1 Cost Per Action (CPA)

CPA allows the advertiser to pay only when the target market performs a specific action on the advertisers website after clicking on a banner on a media site. Examples of applications of this model are site registration or the download of trial version software.

The price of an action is dependant on the industry, product and agreement between the advertiser and publisher.

5.3.2 Cost Per Lead (CPL)

The CPL model allows the purchase/sale of banners based on the cost of each lead generated from the individuals who click on the banner and are brought to the website to request more information. CPL allows advertisers to pay only for the identification of a potential customer who fills in a form to request more information about the product or service.

The price of a lead is dependant on the industry, product and agreement between the advertiser and publisher.

5.3.3 Cost Per Sale (CPS)

The CPS model allows the purchase/sale of banners based on the cost of the amount of sales generated through the website from the individuals who click on the banners.

CPS allows the advertiser to pay only for sales directly as a result of a banner advertising campaign. The industry average CPS is 5 to 30 percent depending on the industry, product and agreement between the advertiser and publisher.

6. Strategy Development

One of the fundamental rules of marketing is to integrate each element of the marketing mix in order to portray a consistent image of the company and its products.

Internet marketing is just one element of the marketing mix, however, it enables communication with the target market in a variety of different ways. The Internet has its own marketing mix, banner advertising being just one element. Unlike traditional marketing channels, the Internet allows the consumer to move seamlessly from each element of marketing communications, clicking from on-line public relations, to banner advertising, to the website all the way to the transaction.

To develop a coherent banner advertising strategy it is important to understand the relationships between each element of the marketing mix in relation to the overall objectives of the marketing strategy. You should be able to answer the following questions for both your on line and off-line marketing strategy.

Marketing Strategy
- What is your marketing strategy?
- What is your aim, objectives and tactics?
- In the long, medium and short term?
- Which elements of the marketing mix are matched to which objectives and tactics.
- What is the relationship between each element of the marketing mix?
- Which brands, products are being pushed the most heavily?

Market Characteristics
- What is the definition of the market?
- What is the size of the market?
- What is the market growth potential?
- What are the key attributes of the market?
- What product and areas are included?

Channel One Banner Advertising Report

Sales
- What are your current sales and market share?
- What are your sales and market share goals?

Target Market
- Who are your key market segments?
- What is their size and segment potential (i.e. no. of individuals in each segment and their expected consumption rate)?
- What are their characteristics, attributes, needs and motivations?
- What are their media habits and shopping behavior?
- What is their attitude towards your company and products?
- What factors impact the decision making process when purchasing products within your segment?
- What percentage of advertising expenditure is spent on each segment?

Products
- What are your key products or services?
- What are their unique factors?
- What are their advantages and limitations?
- What are the differences between your products and your competitors?
- What are your proposed product changes? Why?
- What are the service, warranty, delivery and credit terms?

Pricing
- What are the retail, wholesale prices and profit margins for each product?
- How do these compare to your competitors?
- What are your discount structures?
- Do you have any promotional pricing packages at present?

Distribution
- What is the number and type of your distributors?
- What volume goes through each type of channel?
- What are the key attributes for each distribution channel type?
- Do you have any special agreements with your distributors?
- What are your distributior loyalty programs?

Competition
- Who are your key competitors?
- What are their marketing strategies?
- Which strategies are the most successful? Why?
- How are your competitor's brands perceived by your target market?
- What are the opportunities and threats relative to each competitor?

7. Budgeting

In order to forecast the economic impact of the banner advertising strategy it is important to determine the right metrics for how the strategy will achieve your overall marketing objectives. The first task is to qualify the size of the opportunity for the online marketing of your brand. This is done by answering these questions...

- What is the percentage of your target market online?
- What are the potential sales and economic impact of brand enhancement to this market?

The next step is to determine the cost of converting the potential customer to a sale. This is called the conversion cost. The conversion cost is estimated differently for each banner advertising model.

CPM = CPM X CTR X percentage of customers who will order online.
CPC = CPC X percentage of customers who will order online
CPL = CPL + cost of closing the sale
CPS = CPS + cost of marketing affiliate program divided by estimated number of sales

The percentage of the customers who will order on-line will be directly related to the degree of targeting for your banner. (i.e. how qualified are the users of the site that you will advertise upon? How much do they have in common with your target market?) It is often cheaper to advertise on sites with a low degree of targeting but that will result with a low click through rate. Also sites with a low CPM and high click-through rate generally bring less qualified visitors with a low conversion rate. Therefore you have to do the projections and weigh CPM, CTR and conversion rates to determine the best equation.

The next step is to estimate the impact that banner advertising will have on branding. This is determined by looking at the CPM, or for other models, by estimating the click through and conversion rates and working it back to estimate the number of impressions your banner will have. Then compare the CPM to the estimated dollar value that each impression will have on branding.

This should estimate the return on investment for the banner advertising campaign. Compare it with the conversion costs for other elements of the marketing mix to determine the percentage of the overall advertising budget that should be appropriated to the banner campaign.

Your banner advertising budget should answer the following questions...

- What is the total advertising appropriation?
- What is the total banner advertising appropriation?
- What is the banner advertising appropriation relative to specific target markets?
- What is the total advertising expenditure as a percent of sales?
- What is the estimated return on investment for the proposed expenditures?
- What is the reasoning and evidence to justify the amount of the total advertising appropriation?
- What is likely to happen with an increase/decrease in budget appropriation for each element of the marketing mix?
- What are competitors spending on their banner advertising strategies?
- Is the budget affordable?

8. Targeting

The success of a banner advertising campaign will be directly influenced by the marketers' ability correlate the attributes of the target market with the attributes of users of the website advertised upon.

Channel One Banner Advertising Report

The marketing strategy will have identified the characteristics of the consumers who are most likely to buy the product. The banner strategy will involve the identification of those consumers on-line and the analysis of their Internet usage patterns in order to identify the right website to place the banner.

Banner advertisements should be targeted at specific market segments. Banners targeted with precision produce higher click through rates and have greater effects on branding. Therefore, individual banner campaigns should be developed for key segments such as (primary consumers, business consumers, distributors and market opinion leaders) and segments within those segments.

Each campaign should use the following criteria to match key consumer segments to determine the right websites that they advertise upon…

Demographics

Demographics are factual information about the consumer such as age, sex, income levels, education and geographic location. Demographic information can be used to segment markets and to discriminate buyers and non buyers for a particular product.
Marketers use information from the purchase history of their customers and from surveys to create a profile of the demographic segments with the highest probabilities for purchasing the product.

Advertising websites determine the demographic profiles of their visitors by conducting on-line surveys or by requiring their users to register to the site. Demographic targeting is achieved by matching the demographic profiles of the target market with the profile of the advertising website. It is important to determine the accuracy of demographics profiles on sites without user registration. Many sites use only a small sample of their users for the survey and sometimes the information is out of date.

Sites that require their users to register to the site not only have more accurate demographic profiles but allow targeted impressions to a particular set of demographics (for example, they could target the banner to males over 40 who have an income of over a hundred thousand dollars). However, demographic information can only indirectly predict the consumers needs and motivations. Since demographics are only in-direct, they are therefore less accurate in the prediction of the consumers behavior to the banner and probability of purchasing the product.

Psychographic

Psychographics are information about the consumers' psychological attributes, such as attitudes, needs, motivations, behavior, preferences and lifestyle characteristics. Psychographic information is more useful than demographics for the identification of potential customers and for predicting their response to a banner advertisement. Psychographics directly measure consumer behavior.

Psychographic profiles are compiled from surveys. Psychographic targeting on the web involves matching the profiles of both the advertiser and the publisher. However, this is easier said than done since a large percentage of advertisers and publishers know little about the psycho characteristics of their consumers. Even if they have built an accurate profile, the multitude of variables make it is difficult to find a website that has asked the same psychographic questions as the advertiser.

Firmographic
Firmographics are factual information about a business consumer such as industry category, years in business, number of employees, sales revenues, profits etc. Firmographic information is useful for business to business marketing and is quite accurate for segmenting markets to determine the potential buyers for a specific product.
Firmographics can be extremely useful when websites require users to register their company information since it allows advertisers to target their banner to segments they know are within their market.

Targeting by Content
The Internet is an interactive medium. Unlike traditional broadcasting channels its users are in control of the information that is disseminated to them. They visit sites that contain the content they like and therefore self segment themselves according to their interests.
Demographics are indirectly, but very often not strongly related to customers needs and preferences. Firmographics are more highly correlated with, while psychographics directly measure customers needs and preferences. However, it is difficult and often very expensive to target directly by psychographics and firmographics.

Targeting by content enables the marketer to target indirectly by demographics, psychographics and firmographics. This type of targeting statistically receives a very high correlation with the consumers preferences and needs. This information is not only readily available but it is usually much cheaper than targeting by different types of information.

To target by content markers must develop a profile of the consumer's interests and identify the on-line publications that they read. This process is very similar to advertising in print media, however the marketer must take in to account that further levels of targeting can be achieved within the targeted on-line publications.

Targeting By Activity
The activity that the user of the website is engaged in helps the marketer to understand the characteristics of the consumer and how open that consumer will be to the banner advertisement.

Searching
Purchasing keywords on search engines and directories is an effective way to generate high volumes of qualified traffic to your site. When the user fills in keywords on a search engine related to your product or industry category the banner will be displayed on the search results page. Banners purchased on pages of directories related to a particular category achieve the same results. Banner advertising on search engines will be

determined by your search engine marketing strategy.

Shopping
Purchasing banners on shopping related sites such as product review publications, shopping directories, shopping areas of portals and on shopping malls bring waves of consumers who have a high probability of purchasing your products. It is extremely important to identify the sites where consumers will go to before purchasing your product not only for your banner advertising strategy but also for competitive analysis and for on-line public relations.

Reading
News sites that focused on a geographic region, topic or industry are fast becoming many consumers primary source of information. Visitors to these sites are generally males, from senior management who are likely to purchase on-line and are therefore good people to advertise to.

Researching
Internet users who are focused on researching information on-line may not be open to a banner advertisement even if it is related to their industry. This is because they do not want to become sidetracked from their task at hand.

Learning
Many sites have been developed to educate users about a particular subject. Depending on the situation users can be open to banner advertisements.

Interacting
Sites such as on-line chat rooms where users interact with each other in real-time are generally not a good place to advertise on. This is because the process of interaction has a beginning and an end and users are usually not open outside stimulus during that process.

Targeting by Experience
Consumers use the web in different ways depending on their experience. Forrester Research has placed Internet users into the following categories depending in their experience. Fast Forwards are consumers who use the Internet as a tool for executing a business strategy. Techno-strivers are consumers who are interested in Internet technology. New Age Nurturers generally come from high-income households and Early Adopters are consumers who have been online for a long time and fully understand the medium. Marketers should consider the needs of each category of user and target appropriately.

Targeting by Technology
The Internet provides the marketer with the technology to enable the segmentation of the users of a site along a number of axis. To be more specific, the webserver can tell certain information about the user when their browser requests information from that site. They degree of targeting available will depend on the type of adserver technology used by the site or banner advertising network.

Geographic region
The server can tell the top level domain (.com, .ie .co.uk etc) of the ISP or company

that the consumer is using to access the Internet. Therefore they can tell what country that the consumer is from. This type of segmentation is about 75 percent accurate.

Type of Organization
The top level domain can also tell what type of organization that user is accessing from (i.e. .com is commercial, .ml is military, .edu is educational etc.)

ISP
The domain of the ISP that is used by the consumer to access the Internet can tell a lot about the user and therefore used to target.

Company
Adservers can tell the organization that the user comes from if they access the Internet using a local area network that is connected to the Internet. This enables the marketer to target a particular company or educational institution.

Browser
Adservers can tell which browser and even which version of the browser the consumer uses. Information about users of different browsers can be used to predict certain characteristics; (for example users of a beta version of a new browser are likely to be heavy user of the Internet and open to new Internet related products.)

Operating System
Characteristics of users can be predicted by the operating system and the version of the operating system that they use, (for example, users of a Macintosh are more likely to be in a creative industry such as a graphic design and less likely to be an engineer.

Internet connection speed
New ad server software can determine the connection speed of the user and therefore tell certain characteristics. A major advantage for banner advertising is that it allows the marketer to play multimedia presentations to users of high bandwidth connections who click on a banner.

Plugins
The plugins that are used by the consumer can be detected and therefore used for targeting, (for example consumers who have Realaudio are likely to be interested in music.)

Timing
The time of day can tell a lot about the customer especially when this information is used in conjunction with the geographic region; (for example it can predict if the consumer is accessing the Internet from work or home.)

Placement
The placement of the banner on a webpage and within the site has will heavily influence the click through rate and the effect of the banner on branding. Market research conflicts on the optimum positioning of a banner. The results are subject to the site that is advertised upon. However, Channel One has had the following results from executing our own campaigns...
- banners at the top of pages usually get higher click throughs
- it is best to have your banner on both the top and bottom of the page

Channel One Banner Advertising Report

- banners that appear on the home page have higher click through rates
- but banners on deeper levels of the website have more segmented audiences
- the best place to buy banners are on form results pages (i.e. the page that the user is brought to after filling in a form.)

Frequency

Frequency is the number of times that the customer sees the same banner. Ad servers can measure and determine the amount of times that a banner is displayed to a particular user and even display a variety of different banners in a sequence. Double click was the first company to implement this technology on their ad servers. They found that after the fourth impression click through rates dropped dramatically from 2.7 percent to less than 1percent (this was back in 1996.) They coined this banner burnout, the point that the banner stopped delivering a good return on investment. This study was for a direct response campaign where ROI was measured by click throughs. Banner branding campaigns measure their ROI differently and therefore a higher number of impressions may be needed to get the message through to the target market.

9 Banner Ad Design

9.1 Types of Banners

Banners can be displayed in different media types depending on the advertising network or the site you advertise on.

Standard Graphical Formats

GIF (Graphical Image Format): This is the standard format for displaying images with 256 colors or less on the Web.
GIF 89A: This format allows the animation of GIF images.
JPEG (Joint Pictures Expert Group): This format allows the display of images with up to 16 million colors, it is used for high-resolution graphics such as photos.

Other Formats

Image Map: different areas of the same graphic can be linked to a different page using image maps.

Java: allows the advertiser to run the banner as a small software program or applet. Java banners are approved by the networks/sites on a case by case basis since they take up a large processing time on the client's computer and may compromise the security of the server.

JavaScript: is based on Java but is built into the web browser and allows the banner to perform different interactive applications such as changing the graphic when the mouse is pointed over the banner or by allowing the user to select from a drop down menu.

9.2 Advertising Banner Sizes

Banners come in many different shapes and sizes, the standards below from CASIE and the Internet Advertising Bureau have been adopted by most companies.

Full Banner 468 X 60 Pixels
Full Banner with Vertical Navigation Bar 392 x 72 Pixels
Half Banner 234 x 60 Pixels

Vertical Banner 120 x 240 Pixels
Button 1 120 x 90 Pixels
Button 2 120 x 60 Pixels
Square Button 125 x 125 Pixels
Micro Button 88 x 31 Pixels

The 468x60 banner ad was overwhelmingly the most accepted banner size during 1998. However, the 125x125 and the 120x90 recorded the greatest percentage increase in use during the year. Source AdKnowledge

File Size Limit
Most sites and networks limit the file size of a banner to less than 12K.

Length of Animation
Some networks and sites limit the length of the animation to 4 seconds maximum. A few limit looping which means the animation can be only played once.

Alternative (ALT)Text
Alternative text is the text displayed on the page before the image is loaded. This text pops up when the mouse is put over the image in some of the later browsers. ALT text may be limited to 30 characters.

Tag Text
Tag Text is the text that is displayed under the banner. It is often limited to 30 characters.

9.3 The Message
One of the fundamental rules of marketing is to integrate all elements of the marketing mix in order to portray a consistent image of the company and its products. Therefore the message of the banner ad will be determined by overall the marketing strategy. The message conveyed should be consistent with each element of the marketing mix.

The banner advertisement should grab attention and convey a basic message that effectively relates to the target market through a common motive. Consumers relate to what they can identify with. They should be able to identify with the message of the banner ad and should perceive that message to be of relevance, interest or value. The message should be focused on the objectives of the marketing strategy, not just on generating click throughs. It should qualify the target market before they click on the banner and therefore drive a highly targeted audience with a high purchase probability to the website.

The message should contain an offer that is simple, stated clearly and obvious to the user. The consumer should understand how easy it is to take advantage of the offer. The offer should be in no way confusing or overly elaborate. Cryptic messages may generate a 18 percent higher click through rate but they don't work because the consumers who click are not properly qualified and the banner has no impact on those who don't click.
The following techniques have been found to increase click through rates and the impact that the banner has on branding.

Ask questions
A question in banner messages such as, Where do you want to go today? How many Internet users are on-line? Where can you find out everything you need to know about banner advertising? Initiate

interaction with the consumer thereby increasing the effectiveness of your banner.

User Interaction
Advertisements that allow the user to do something straight away are extremely effective. Examples include, Download a free evaluation now, Search for the web here, Find a gift for your loved one now!

Free offers and incentives
Free offers and incentives catch the eye of the consumer and increase click through rates.

Competitions
Ads that give the consumer a chance to win something can generate phenomenal click through rates.

Sense of urgency
If the offer has a sense of urgency to it, (for example, Free access today only), consumers are more likely to click on the banner since they realize the opportunity cost of not clicking.

Call to action
Messages containing phrases such as Click here, Enter Now, Sign up Today all have a positive impact on click through rates.

9.4 The Website
The page that the consumer is brought to after clicking on the banner should be consistent with the message, the offer and the needs and preferences of the targeted segment of the market.

9.5 The Design
There are 2 strategies for designing banners. The first is to make the banner jump off the screen and grab the consumer's attention. The second is to integrate the banner with the design of the publishers so that the consumer thinks that they are clicking on a link within or associated with that site.

Animation catches the users eye and generally grabs attention more effectively than still banners. Integrating features of the personal computer such as mouse pointers, status bars and buttons have a positive effect on click through rates. Bright colors such as yellow, blue and green grab attention more than whites, blacks and grays.

Executive Summary: The State of One to One Online

Title: *Executive Summary: The State of One to One Online*
Author: *Peppers and Rogers Group*
Abstract: *We found that 1 to 1 marketing online has advanced slowly since our first report in April 1999, and there is still a long way to go. Financial Services sites scored highest of all industry categories for several reasons. First, they provide customers with Internet-based, easily digitized financial information and tools to manage their portfolios.*

Copyright: Peppers and Rogers Group

Any one-to-one Web site is based on the simple idea of treating different customers differently. Our new framework includes the basic implementation steps of one-to-one marketing: customer identification, customer differentiation, customer interaction and product or service customization. Readers familiar with Peppers and Rogers Group recognize this IDIC framework, since it is at the heart of what is required for building an effective 1to1 program within a company. (Please see our Web site at www.1to1.com for more information about this framework). Using the IDIC framework, we placed the capabilities or features required of a Web site into one of the four dimensions. For example, sites that effectively identify their customers must provide privacy assurances.

We reviewed more than 150 1to1 Web sites from around the world, culled from lists of "top sites" from multiple sources, and supplemented with recommendations by readers of our first report and Peppers and Rogers Group consultants. Each site was first screened to see whether it had any personalization features and recognized return visitors. We ranked the remaining 150 sites and selected the top 65 for extensive review. We then placed these 65 sites in one of eight categories or industries:

- Books/Music/Video (e.g., Amazon)
- Community/Calendar (e.g., When)
- Consumer Goods (e.g., Gap)
- Consumer/Business Services (e.g., FedEx)
- Financial Services (e.g., S&P Personal Wealth)
- Information Technology (e.g., Dell)
- Portal/Media (e.g., Wall Street Journal)
- Travel/Hospitality (e.g., Biztravel)

Key Findings

Overall, we found that 1to1 marketing online has advanced slowly since our first report in April 1999, and there is still a long way to go. Financial Services sites scored highest of all industry categories for several reasons. First, they provide customers with Internet-based, easily

Executive Summary: The State of One to One Online

digitized financial information and tools to manage their portfolios.

Since the Internet is a natural medium for Financial Services, there are literally thousands of competitors from which customers can choose. This "hyper" competition typically drives a market quickly to a mature state, and we found most of the Financial Services sites quite sophisticated relative to other sites in their deployment of 1to1 marketing techniques.

While there were many differences in the degree to which different sites and industry groupings used 1to1 marketing online, there were also some striking similarities across all four IDIC dimensions.

IDENTIFY

Most sites respect customer privacy...
Almost every reviewed site had a well-stated privacy policy that reassured customers as to how their information would be handled. The sites included a statement about how the data would be used, and most gave customers the option to prohibit the sharing of their information with others. About half of the sites were members of a third party privacy program (e.g., TRUSTe), an added assurance for customers about the use of their data.

...and offer enticements to register...
Nearly all of the sites use enticements to get users to provide personal or company information. The most common enticement was access to more functionality. Not as common, however, was an explicit statement of what a customer would get in return for sharing information (e.g., more functionality, better features).

...but are still learning how to gather the data...
About half of the sites are using a form of "drip irrigation," or limited questioning, to gather customer data (see section on drip irrigation for further explanation). Customers can be impatient when divulging information and will leave if the process is too long or burdensome.

DIFFERENTIATE

All top 1to1 sites recognize returning customers...
All of the sites reviewed recognize returning customers with some personalized greeting. Most of them do it without "cookies," or small programs left on the customer's computer at registration. Most sites also allow customers to override cookies.

...many are organized around customer needs...
More than half (60%) of the sites are organized around customer needs or some type of segmentation (e.g., enterprise, small office). This helps customers feel their needs are understood and being addressed.

...but lack differentiated customer support...
Surprisingly, most sites offer every customer the same basic level of support online. This is contrary to good customer care programs that recognize premium

Executive Summary: The State of One to One Online

customers and offer them additional support or access channels.

...and rarely incorporate data from other parts of the enterprise into the Web experience...
Most sites fail to link their back-office systems and related customer information to the Web experience. Indeed, only eight percent of the sites reviewed appeared to link their front-end and back-end systems. Internet-only businesses, however, were an exception.

INTERACT

All sites support communications via email....
All of the sites reviewed support customer inquiries via email. Only about half, however, offer a response the same business day. About half the sites also "push" communications to their customers via email in the form of newsletters or notices.

...and offer online support databases...
All sites offer some kind of self-service customer support database for answering basic questions. About half of these databases are searchable.

...while most sell products and services over the Internet but still lack "1-click" ordering...
Two-thirds of the sites sell products or services over the Internet. Just under half of these allow customers to view their order history online and offer online order tracking. Only a few of the sites offer "1-Click" ordering, which makes the ordering process so much faster by "remembering" what you have previously told the company about your billing or ship-ping preferences.

CUSTOMIZE

All sites personalize the Web experience to some degree...
All sites offer some personalization for customers. Three-quarters of the sites address customers by username. Only one-fifth offer personalized wish lists and product recommendations. A few of the sites provide personalized promotions or coupons.

...and allow their customers to customize the site...
Most sites allow customers to customize the look and feel of the Web site. This may be simply the layout of information or the kind of information displayed.

...while most use partners to offer a broader range of products and services...
Two-thirds of the sites partner with other companies to provide a broader range of products and services. Only about half of these partnerships are well-integrated into the site, however.

Web-based Sales: Defining the Cognitive Buyer

Title: Web-based Sales: Defining the Cognitive Buyer
Author: Paul Zellweger
Abstract: Marketing on the Web represents a radical departure from the traditional thinking. Traditional mass marketing channels broadcast information from as single source to mass audience. The relationship represents a one-to-many communications model where the entire audience is treated the same way. This article explores the forces contributing to the transformation of buyer behavior when buyers use the Web for making purchase decisions. These forces include trends in the marketplace, where products are becoming more complex and plentiful.

Copyright: ArborWay Electronic Publishing Inc.

Abstract

This paper explores the forces contributing to the transformation of buyer behavior when buyers use the World Wide Web for making purchase decisions. These forces include trends in the marketplace, where products are becoming more complex and plentiful, as well as the emergence of Web-based marketing and functions. The interactive nature of electronic sales, like Web-based marketing, makes this type of sales significantly different from its traditional counterparts. Sales on the Web shape buyers' perceptions about products, the decision making process, and the marketplace, resulting in buyer behavior this paper defines as the Cognitive Buyer. Technical developments like secured transactions and improved access methods like catalog content menus will address buyers' uncertainties and provide marketers with a better understanding of their buyers. As this new sales channel develops it will create a demand for itself by using knowledge systems to organize information.

"Focusing on mere information has led to overload...rather than the search for meaningful new patterns of knowledge"
Hazel Henderson

Overview

Buyer enthusiasm for direct sales on the World Wide Web lags far behind industry expectations and the questions being raised all seem to miss a larger transformation taking place. The combination of changes in the marketplace and the distinctive nature of the electronic sales channel all contribute to the emergence of a new and distinctive buyer behavior that this paper identifies as the Cognitive Buyer. The Cognitive Buyer relies heavily on rational problem solving and abstract reasoning and differs from traditional buyer behavior by intentionally engaging technology in the decision-making process. This paper explores how this technology is shaping buyer behavior in order to understand why sales on the Web have been so disappointing.

Web-based Sales: Defining the Cognitive Buyer

Technology is employed in all aspects of product development and marketing, and most recently in attempts to automate sales. Bloch, Pigneur, and Segev (1996) present a fundamental business rationale for using electronic commerce derived from its potential capabilities to improve, transform, or refine current products, process or business models. Hoffman, Novak, and Chatterjee (1996) identify six different commercial opportunities on the Web by function, including the online store or electronic product catalog, the primary focus of this paper.

Sales applications on the Web have evolved in a very short time and progressed through easily identified stages of development. First, there were flat product brochures. Next, came the current offering of interactive electronic catalogs that enable buyers to search databases of products to find what they need. Yet, only a handful of success stories exist and, most recently, IBM closed its electronic mall, and Nets Inc. in Cambridge, Massachusetts declared bankruptcy. Cronin (1997) argues that the demand for online sales on the Web can only come from customers, competitors, or universally accepted distribution channels.

Transformations in the Marketplace and in Marketing

To understand the influences affecting buyer behavior on the Web one only has to direct his or her attention to two major changes occurring in the marketplace. First, technology has dramatically reduced the amount of time it takes to develop a new product and bring it to market. This has set off increased competition and produced an extraordinary array of product choices in both the consumer and business markets. Aggressive product marketing and rapid innovation make these markets very dynamic and fast paced.

Second, the increased use of computer technology in the products themselves has expanded product capabilities and made products more feature-rich and complex. The overwhelming amount of detailed product information is thrust directly on the buyer and forces buyers to take a more deliberate and studied approach when making a purchase.

Traditional marketing and media channels also are feeling the impact of changes in the marketplace and are losing their effectiveness to reach individual consumers. Rayport and Sivoka (1995) describe this market condition in more basic terms as an "overcapacity, in which demand, not supply is scarce". The Web is a logical alternative to traditional marketing because it is technically capable of addressing these issues, including the mounting flow of detailed information and reaching out to individuals. However, aspects of its interactive capabilities are not fully utilized and customers are, for the most part, anonymous and freely able to move from one site to another. Overall, Web marketing has yet to identify issues related to creating demand, other than aspects of personalization where promotional materials respond directly to individual buyer's interests.

Hoffman and Novak (1996) argue that marketing on the Web represents a radical

Web-based Sales: Defining the Cognitive Buyer

departure from the traditional marketing. Traditional mass marketing channels broadcast information from a single source to a mass audience. The relationship represents a one-to-many communications model where the entire audience is treated the same way. On the Web, access is interactive and information flows between buyer and seller. This rep-resents a far more complex set of relationships and a different communications model, many-to-many, that includes the means to tailor a response to an individual need. The primary challenge for Web marketers is to find new ways to develop interactive capabilities that can broaden its appeal as a mass marketing channel.

Up to now, the merger of marketing and technology on the Web has produced its own unique form of personalization based on a buyer's navigation activities or information requests. Marketers use this information to sense or anticipate buyer behavior and respond to it by presenting ads and promotional materials directly related to the buyer's interest.

Conceptually, the process creates a one-to-one communication between a buyer and a seller. However, a central problem with personalization on the Web, and Web-based marketing in general, is that neither of these activities generates a sufficient demand for sales.

Forces Transforming Buyer Behavior

To observe how market forces have initiated changes in buyer behavior, simply look at the PC hardware and software market in the last ten years. This particular market has always moved fast and provided buyers with a wide selection of competing products. Its products are also somewhat different as they are technically-based and complex, and not marketed exclusively through consumer or business market channels.

As more buyers overlap the traditional consumer and business markets product marketing and distribution has also changed. Large warehouse stores like Costco and BJ's target sophisticated technology products to both small businesses and consumers. Today it is not unusual to see a software product or, even an Intel microprocessor, advertised on television or in popular magazines.

In other, more dramatic ways the frequent rate of innovation in the PC market has prepared the buyer to anticipate product change. Buyers now accept the fact that any product they buy for their computer today may be surpassed by something newer, better, or cheaper tomorrow. This type of innovation not only shortens the expected lifespan of a product; it also compels buyers to consider market trends before making a purchase. In this

Figure 1: Buyer decision process

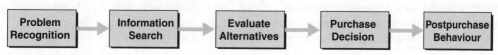

environment, product information, particularly current and scheduled features, takes on a more direct role in creating demand for a product or avoiding purchase hesitation.

reconsider his or her original problem. Each time this happens Problem Recognition is revisited and the overall effect transforms the sequence of events in the decision process (see Figure 2.).

Figure 2: Dynamic buyer decision process

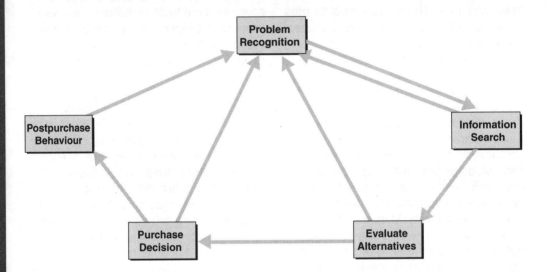

The buyer's experience in the PC marketplace closely foreshadows the speed and structure of buyer behavior on the Web. In a traditional buyer decision process (Kotler 1988), a consumer progresses in a linear fashion from a Problem Recognition state to an endpoint, the Postpurchase state (see Figure 1). Yet, in a very dynamic market, like the PC market, the decision process is less predictable and often more erratic. Buyers can come across pertinent information when doing other things and new product information is constantly being released. This includes innovative ways of doing things or new perspectives on product usage that cause a buyer to

Adding sales capabilities to the Web challenges not only the traditional ways of analyzing consumer buyer behavior; it also challenges the structure and substance of the consumer market. For instance, Web-based technologies like intelligent shopping agents threaten to turn every product into a commodity simply by automating a search that can find the cheapest price. This type of automation, along with its general purpose search engines, sets this market apart from all others. No other market gives the buyer as much freedom and control, or can deliver as many resources that can be applied to the purchase decision.

Web-based Sales: Defining the Cognitive Buyer

The Emergence of the Cognitive Buyer

Web-based sales will have more impact in shaping buyer behavior than any other sales channel. Yet, little attention has been directed towards the buyer, in large part due to the short time online sales have been available. To ignore the buyer at this point would be a grave mistake, because the discussion will remain, as it has been almost from the start, unbalanced and limited to buyer benefits (convenience, searching capabilities, price comparisons, control, and so on). By not taking the buyer's perspective into account the discussion can only set false expectations and miss the more substantive issues facing this newly invented exchange between buyers and sellers. Not surprisingly then, buyers have been slow to respond to sales opportunities on the Web.

Meanwhile, many explanations have been proposed to overcome uncertainties about the Web's potential as a marketing and sales channel. Some, like the one presented by Regis McKenna (1995), direct attention to the interactive capabilities of the Web and how they can be used to draw customers into a marketing conversation. Other, more divergent ones, address issues related to the buyer's perceptions about the shopping experience or security. It should be noted that these points all have merit but somehow fail to acknowledge the larger transformation taking place, namely the influence of market forces and technology on buyer behavior. Web-based marketing will shape buyers' perceptions of what constitutes a product. As products get more complex and difficult to use they require more pre- and post- sales support to help buyers understand what they need to know. The Web is ideally suited to deliver these services and respond to prospective buyers' concerns by answering questions and making this information available. Buyers will come to expect these services, particularly when it is so easy to inspect a product and its support on the Web and compare it with others.

The Web also can respond to the buyer's need for other types of information like news and analysis, and it can provide entertaining ways for learning about new things. This will attract the type of buyers who fit the Cognitive Buyer style profile. These buyers have directly benefited from previous breakthroughs and have developed a positive attitude towards technological and change. They are optimistic, open to change, and welcome new ways of doing things that save time or money or both.

Cognitive Buyers are well educated and have sophisticated tastes. When these buyers make personal purchases they are more likely to engage in a form of complex consumer behavior revealed by postmodern marketing research, behavior characterized by, among things, people who do not always remain true to their predicted type (Ogilvy 1990). These are buyers who are highly aware of their motivations and the motivations of others. In a rapidly changing world, the Cognitive Buyers will seek out products and services that will help them adapt to change.

Web-based Sales: Defining the Cognitive Buyer

Direct Sales on the Web

In essence, sales on the Web are as different from traditional sales as Web marketing is from traditional marketing. The Web's interactive capabilities, global presence, and abundance of "live" information are all unique. A prospective buyer has extraordinary access to a wide range of sales materials including pricing, support, polices, competitive analysis, and more, all in one place. These features give the buyer an unprecedented level of control in a sales setting where there is virtually no pressure to buy.

Sales professionals intuitively know that to make a sale, they have to reduce the buyer's perception of risk and uncertainty. In this respect electronic sales are no different from any other type of sales channel. In fact, the novelty and uniqueness of cyberspace can only add to an underlying concern that occurs to any buyer in an unfamiliar context. Yet, there is something more fundamental that keeps buyers away— the perception of a non-secure environment.

Ironically, the credit card companies themselves seem to contribute to this fear by not sending a clear message that they are able to protect buyers from credit card fraud. In theory, direct sales on the Web make perfect sense, but on a practical level there is wide range of management problems, both internal and external, that make this new sales channel far more complex and different than buyers and sellers previously thought.

Even from a technical standpoint, the search engines, used to locate information on the Web, contribute to a collective uncertainty about the environment. No two search engines produce the same results. The problem is these systems are based on "word matching" methods, an information age technology whose limitations are well known and easily recognized. Searches generate too much unnecessary information and they can simultaneously miss something important because the right keyword was not used.

Furthermore, most catalogs on the Web, no matter how big or small, assume the buyer knows what he or she wants and how he or she plans to find it. Smaller catalogs typically use a table of contents that links to a category of products that, in turn, is linked to a set of product pages. For the buyer, navigating from one page to another can be extremely frustrating, especially when pages contain irrelevant information.

With larger product catalogs the access problem only worsens, as database retrieval technologies provide little or no direct information about the products they fetch. When menu-based attribute/value searches and parametric search strategies are employed, the technology improves matters for buyers familiar with the content. Yet, a central problem with these methods is that they do a poor job of encouraging a buyer to browse or, worse, they do not provide any browsing capabilities at all.

Web-based Sales: Defining the Cognitive Buyer

Responding to the Buyer's Needs

The Cognitive Buyer has an appetite for information and details, and Web technology can directly respond to this interest. When you consider information in its broadest terms — "that which reduces uncertainty" (Shannon and Weaver 1949) — this medium is ideally suited to helping buyers make purchasing decisions. But sellers should never underestimate a buyer's need for information or the extent of his or her caution. Since ancient times the notion of caveat emptor — let the buyer beware — has made it perfectly clear that it is the buyer who is responsible for gathering information and analyzing the risk.

As noted above, buyers rely on information to predict favorable outcomes and minimize risk. In a supermarket a buyer will squeeze a piece of fruit to make sure it is not overripe. Touch yields valuable information about the condition of the fruit. So does its appearance. The vendor, seeing the buyer with a piece of fruit in her hand, also may try to reassure her about its freshness by telling her when it arrived and where it came from. In an abstract context like an electronic market or a paper-based catalog, the medium can accomplish the same objective as the fruit vendor by carefully providing the right mix of information.

A buyer uses perceptual processes, like touch and fit, to get information. The perceptual processes themselves follow a clear developmental progression from visual-motor skills like vision and touch, to visual-auditory skills like vision and hearing, to visual-cognitive skills associated with reading and writing. Higher-level perceptual processes can substitute lower ones, and still communicate the essence of the original information but in a more filtered, abstract way. For instance, you can get an impression about a place by looking at pictures or hearing a story without ever being there. And a combination of perceptual processes (hearing music, looking at film clips, and reading about it) goes a long way to enrich the information and strengthen the message.

A sale on the Web will never be able to compete with a direct sales experience. Shopping is an interpersonal experience created by a salesperson and the ability to inspect products by holding or touching them. Using a catalog to make a purchase represents a more abstract exchange that obviously does not have the same emotional pull. Yet the convenience and hassle-free aspects of Web-based sales are clearly benefits. For example, for a buyer who does not have the time to go to a local fruit stand to shop, a virtual fruit market on the Web would make perfect sense. That is, if the vendor understands how to market "freshness" on the Web. Obviously, this would include an inviting picture of the fruit and a description of it and how it would be shipped to the home. In addition, the seller would have to post an aggressive customer policy to address any doubt in the buyer's mind, like "The Strongest Guarantee in the Business" used by the highly successful catalog fruit vendor, "Harry and David" (see www.harrydavid.com).

Web-based Sales: Defining the Cognitive Buyer

The Web's interactive capability has the potential to approximate the dialogue one would hear between a buyer and seller. Therefore it is important to note that a good sales professional always acts on several assumptions that he or she knows will improve the likelihood of making a sale. This includes acknowledging differences among buyers and the need to answer questions and make suggestions. To achieve these goals sales professionals carefully listen to the buyer's questions to identify individual needs and respond to questions in a careful and deliberate way. The question/answer exchange forms the basis for a dialogue that addresses uncertainties in the buyer's mind and enables him or her to be more confident about the purchase decision.

Creating Demand by Reaching Out to the Buyer

As the Web organizes product information and delivers more customer support services, it creates a demand for itself. Prospective buyers currently turn to the Web to gather and compare product information. For these buyers the Web has a gestalt-like effect where its perception as a resource for sales materials exceeds the sum of its parts.

Companies are motivated to attract Cognitive Buyers to their sites on the Web because they are sophisticated buyers and potential product champions. By engaging these buyers in Web marketing, a company will enable a two-way knowledge transfer where these buyers can learn about its products and the company can learn from its customers.

McKennan (1996) describes aspects of this knowledge exchange as drawing the customers into a marketing conversation that helps the company develop relationship marketing, prepare markets for new products, and define future trends. Deighton (1996) points to two critical features of this conversation— the ability "to address an individual" and remember his or her response —as aspects of the exchange that are vital to transforming Web-based marketing into a good conversation with the customer.

While the Cognitive Buyer represents a specific type of buyer behavior, defined by his or her use of the Web, it is important to note that these buyers are not a homogeneous group. Differences among these buyers can be identified by differences in their knowledge or cognitive understanding of a particular product market and its products. For example, buyers of home VCRs or molecular biology restriction enzymes both face the daunting task of knowing the marketplace, the products in the market, and features associated with those products.

In personal sales exchanges, buyers turn to sales professionals to help them learn about product benefits and understand product choices. For the Web to provide this type of information more sophisticated access methods would have to be used. Alternative methods, derived from knowledge based tools, are better able to help buyers understand product differences and engage buyers in a mutually beneficial way. However, traditional technologies such as artificial intelligence and expert systems that could help companies answer

Web-based Sales: Defining the Cognitive Buyer

buyer's questions and provide this type of information such far too costly and present too many risks.

Understanding the Cognitive Buyer

With the recent introduction of content menus to the Web (Zellweger 1997) companies will now be able to use their product catalog to engage their buyers in a highly efficient marketing conversation that helps both buyer and seller. Catalog content menus organize product access according to buyers' needs, highlighting important product details, benefits, and functionality. Buyers use the catalog's menu lists, like an index in the back of a book, to guide them to relevant products. The menu technology is derived from an underlying knowledge base that provides unlimited cross-reference capabilities.

Buyers navigate through successive menu paths to locate answers to their questions and marketers record these paths to monitor buyer behavior and identify buyer preferences.

To identify differences among buyers, marketers first identify the cognitive associations buyers make to locate a product through a process called knowledge segmentation. This starts out with a general understanding of a specific marketplace like home electronics or molecular biology products and includes categories of products such as VCRs, tape players, and video cameras or cloning vectors, kits, and restriction enzymes. The next level or segment represents a knowledge of specific products within these product categories. And finally, the last segment or knowledge level represents a detailed understanding of individual products, including benefits associated with specific features, attributes, and capabilities.

Knowledge segments help marketers acknowledge differences in buyers' understanding of their products and the market itself. These differences correspond to novice, intermediate, and expert product skill levels. The hierarchical ordering of segments helps a content expert design a succession of menus that correspond to the progression used by a salesmen to help a buyer articulate needs, narrow the field, and make a purchase. Novices can select the most general topics and progress towards more specific details until a product matches their need (see Figure 3). Intermediates and experts have the option of using more direct routes according to their product skill level.

Figure 3: Knowledge segmentation

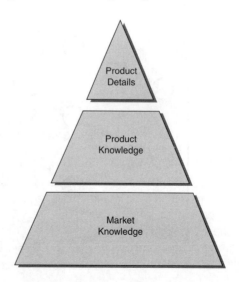

Marketers can also use a catalog content menu to conduct market research to identify differences within the same knowledge group. This includes monitoring changes in product usage and the buying patterns of special groups. For instance, the biotech industry is multidisciplinary and different scientific disciplines often use their own terms to identify the same thing. The marketer creates special menu paths for each discipline and then can monitor these menus to track each groups' usage.

By enabling access to the widest possible range of buyers, content menus establish yet another communications model, many-to-one (see Figure 4). In this model the seller provides multiple access paths to an individual buyer to assure the buyer that the seller understands his or her needs and can answer his or her questions. In more far reaching ways, the many-to-one communications model provides both a rationale and a framework for developing the Web's mass marketing capabilities.

Conclusion

The explosion of products and technology has created an exciting opportunity for automated sales on the World Wide Web. Yet, Web-based sales is still in the early stages of demonstrating its potential as companies grapple with issues ranging from sales management, marketing, technology, and even the law. For the buyer, sales on the Web still present a major risk and uncertainty, due in large part to a lack of secured transactions.

Figure 4: Many-to-one communications model

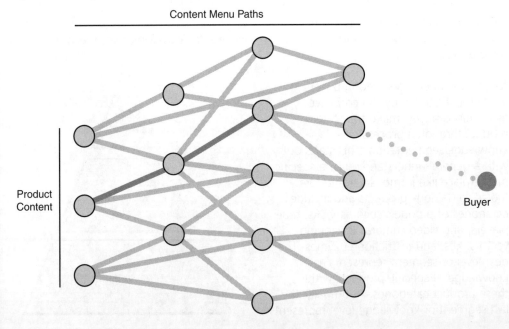

Web-based Sales: Defining the Cognitive Buyer

This paper argues that sales on the Web represent a radical departure from traditional sales channels. On the Web, product marketing and sales merge and create a unique marketplace that challenges our traditional ways of analyzing buyer behavior. What emerges from these developments is a distinctive buyer behavior, identified as the Cognitive Buyer, that uses technology to automate aspects of the decision making process.

The Cognitive Buyer recognizes the Web as an information resource, as well as the means to purchase and support products. The medium's effect on the buyer creates a gestalt where the sum of its parts far exceeds its identity as a marketplace. As more product information and services become available on the Web, the Web will create a demand for itself as an alternative to traditional sales channels.

The Web's potential as a mass marketing channel is far more complex than earlier thought. When information age access methods are applied to database searches of product catalogs these methods restrict access and hide content from buyers. Yet, when companies use knowledgebase methods they create a new communications model, many-to-one, that enables buyers to browse and explore product catalogs. It also enables knowledge to flow in both directions, helping buyer and seller. This type of communication creates a dialogue that will pave the way for mass marketing on the Web by using knowledge to organize and mediate information.

Future Research

This paper briefly looked at changes in the marketplace to understand how the purchase decision process has been transformed by Web technology. Other conceptual differences in buyer behavior involvement between the Cognitive Buyer and the consumer, identified by Foxall (1992) as the Cognitive consumer, exist and should be considered in any subsequent discussions about these buyers.

Buyer behavior on the Web also will be greatly influenced by other mass marketing channels like advertising on television. Companies currently use this advertising build a brand awareness and attract buyers to their Web sites. With the eventual integration of television broadcasting and Web access the impact on buyers may reveal even more varied types of buyer behavior and styles.

Directed research on access methods like content menus and buyer's navigation patterns will most likely provide the most fruitful way to study how these forces will influence buyer behavior and generate different styles. Perhaps the most exciting aspect about these research methods is that they are essentially part of larger marketing and sales systems that can provide empirical results in real time. These developments will expand marketing research capabilities to a larger number of smaller companies and may have as much an effect on the development of Web marketing and sales as the study of technology-enhanced buyer behavior.

Interactive Relationship Marketing

Title: Interactive Relationship Marketing
Author: David M. Raab
Abstract: It is an embarrassingly common observation that the Internet is well suited for relationship marketing. But it's less clear which technologies you need to make it happen. What type of database is suited? And note this: the database is much less important in the interactive world than the interactive application itself. So just what are those interactive applications anyway?

Copyright: Raab Associates © 1998. All rights reserved.
Biography: David M. Raab is a partner in Raab Associates, a consultancy specializing in marketing technology selection and evaluation. He is author of the Guide to Database Marketing Systems and can be reached at draab@raabassociates.com.

By now, it is embarrassingly common observation that the Internet is well suited for relationship marketing. What's less obvious is exactly which technologies you need to make it happen.

Start with the database. Surely this must be the foundation, just as it has been for non-Internet forms of relationship and database marketing? And surely it will look like the databases developed for these other applications?

Think again. Conventional marketing databases are designed for segmentation and analysis, not to execute real-time interactions. Analytical applications require data structures that make it easy to ask complicated questions and to scan large amounts of data quickly. These typically involve some type of "denormalized" design, of which the best known is the "star schema" popularized by data warehouse applications. These structures are quite distinct from the "normalized" designs used in conventional transaction processing systems, which are very fast at finding and updating records relating to a single account.

("Normalized" structures store each piece of information only once, which means elements must be grouped into many different layers or tables; "denormalized" structures use a smaller number of tables but accept some redundancy. For example, a normalized design would have one table for household information such as street address and a separate table for information about individuals. A denormalized design would combine these into a single table of individual records, with the address information repeated on each record. The denormalized structure is easier to query, since there is no need to connect household records to individual records. But the denormalized structure is also harder to update, since several records must be changed to record a new address. This takes more time and leaves considerable room for error.)

Interactive Relationship Marketing

One of database marketing's grandest crusades over the past fifteen years has been to convince the larger technology community that marketers really do need a separate database with a different type of structure. This battle has finally been won, more because the much more powerful data warehouse developers had similar needs than because of anything the marketers themselves really accomplished. But now, in the sort of irony that is only amusing to people who haven't spent a decade in the trenches, interactive marketing does require quick access and updates to individual records. That is, it requires a database using a conventional, transaction-oriented structure, not the finally-accepted separate marketing database.

Just to clarify matters--or possibly confuse them further: interactive marketers still do need their consolidated, denormalized, analytical databases as well. They will use these databases to find behavior patterns and market segments, to evaluate results, and to define the parameters of new campaigns. These are the traditional analytical uses of a marketing database. But non-interactive marketers also use this same database for their operational function, which is to select names to receive promotions. The analytical database can handle this operational application effectively, because its selections are made against the file as a whole in some sort of a batch process. But the operational application of the interactive marketers involves reading single records as the transactions occur. This is not something the analytical structure handles well. This is why the interactive marketers need a separate database for their marketing operations.

Another way to express this is to say that for interactive marketing, the focus shifts away from database to the application. This is a radical change in perspective from conventional database marketing, where building and maintaining the consolidated marketing database has always been the first and greatest challenge. (In fact, one of the increasingly common issues faced by conventional database marketers is figuring out exactly what to do with their centralized database now that they have finally managed to build it. This need for efficient strategy definition and implementation is driving much of the industry's current software development. But that's a topic for another day.)

The application-centric view of the interactive marketing systems is illustrated starkly by their ability to function without any external data at all. Many of the interactive systems are designed to make marketing decisions based solely on the behavior they observe during a visitor's current session in the Web site. Of course, they prefer to keep a history of behavior over multiple sessions, but recognize the technology does not always allow the system to identify the same person from one session to the next. Even systems that do rely on tracking behavior over time are not necessarily able to integrate information gathered at the Web site with non-Web sources such as a conventional marketing database.

Part of the reason is technical--a visitor can often access a Web site without providing

Interactive Relationship Marketing

a name, account number or other identifier that will link to other company systems. But many systems ask visitors to register, thus providing information that could potentially be used to match against other sources. Even those system often choose instead to build their own internal database by gathering additional profile information directly during the registration process. Using an internal database makes it easier to design and operate these systems, since they have full control over the format and structure of the data they must access. It also involves an implicit judgement that these advantages outweigh the benefits of integrating with external data. Coming down on the side of efficiency over integration is the classic choice of operational application developers--and quite foreign to the traditional marketing goal of a single, consolidated database.

So even the level of the database itself, interactive marketing systems are profoundly different from conventional database marketing operations. Not only do they use a different data structure--operational rather than analytical--but they are less focused on building integrated databases than on using whatever data is available to manage interactions. Of course there are exceptions to the rule: products including Persimmon IT TargIT (800-546-7242; http://www.persimmon.com/) can in fact access external databases directly. But most others rely on data they gather themselves or, at best, import from external systems through a batch process. There is no small irony here: the most powerful systems for relationship marketing cannot easily access the full history of a customer relationship. But the database is not the only technology involved in interactive relationship marketing.

The marketing database, which plays a central role in conventional relationship management systems, is much less important in the interactive world than the interactive application itself. So just what are those interactive applications, anyway?

In broadest terms, these products generate responses that are tailored to the actions of Web site visitors. This involves identifying the set of possible actions, creating a list of possible responses, and defining rules that determine which response goes with which action.

Products differ in how they handle each of these capabilities. The "actions" they consider may be limited only to events during the current Web session, or may include a richer context provided by actions during previous sessions, information the visitor has provided through Web registration or surveys, and information derived from non-Web sources such as a marketing database. Many products use this information to place the visitor in a market segment and then apply different rules to different segments.

Due to the quasi-anonymous nature of Web contacts, it is a significant technical challenge to link a visitor with any information beyond the current Web session. The most direct solution for visitors to identify themselves through registration or use of an existing identifier such as an account number. This makes it easy to store and retrieve their information and probably makes the most sense in the context of an open, two-way marketing relationship. Nearly every interactive relationship system can work with registration data.

A more passive alternative involves placing a "cookie" on the visitor's PC. This is a small file that stores a record of past behavior at the site without requiring specific action by the visitor. However, cookies are problematic for several technical and privacy reasons. A more acceptable Open Profiling Standard is being developed by the World Wide Web Consortium's Privacy Preferences Project (commonly referred to as P3P). Details on P3P can be found at http://www.w3.org/.

A key element of the Open Profiling Standard is that it will be common across applications--potentially allowing different Web site owners to access the same visitor information. Developers including Engage Technologies (978-684-3884; http://www.engagetech.com/) and Firefly (617-528-1000; http://www.firefly.net/) have already developed their own "portable" profiles, which are intended to serve cooperative networks of sites that will use them to improve ad and message targeting. These profiles use standard formats and common survey questions, so information can be accessed by any interactive application. The information is encrypted to prevent outsiders from viewing it and users are supposed to consent to their participation. Microsoft recently purchased Firefly for this technology.

Linking Web-generated information with external files such as a marketing database requires either a common ID such as an account number or a name/address match similar to a conventional data consolidation process. Interactive marketing products generally leave any advanced matching to other systems, although many can import external data given an ID to match against. Persimmon IT TargIT (800-546-7242; http://www.persimmon.com/) is unusual in its ability to maintain a live linkage between an external relational database and the Web application itself. This lets it assign segment codes in the external database and look these up as visitors enter the Web site. The more usual approach is to import the external data, generate the segment code, and store it on the user profile in the interactive application's internal database.

The interactive systems also vary considerably in the rules they apply to their data. Most allow users to set up a hierarchy of rules use a combination of user actions and profile data to determine which content will be provided. Products like Intelligent Interactions Adfinity (703-706-9500; http://www.ipe.com/), with roots in selecting banner ads, add extensive features to control when, how often and in what sequence each visitor sees different messages. Rubric EMA (650-513-3870; http://www.rubricsoft.com/) and MarketFirst (408-261-6950, http://www.marketfirst.com/) can include complex campaign flows, with branching sequences of actions and options for tasks like report generation and sending messages (such as sales leads) to people other than the visitor. Persimmon IT can

Interactive Relationship Marketing

also execute this sort of complex marketing campaign, and even offers the type of champion/challenger testing and detailed response analysis used by conventional database marketers. This adherence to standard database marketing procedures is quite rare among interactive systems, although it will eventually be necessary for properly controlled operations. Most of the interactive products (and their users) have not yet reached the state of maturity where they realize they need it.

One of the greatest challenges in setting up rules for Web-based interactions is sheer volume: the number of options and rate of change are so high that manually defining an optimal set of rules is impractical, and probably impossible. Aptex SelectCast (619-646-0121, http://www.aptex.com/) and Nestor InterSite (401-331-9640; http://www.nestor.com/) both use neural network models in combination with more conventional rules to determine which messages are most appropriate for a visitor. Somewhat similarly, NetPerceptions GroupLens (612-903-9424; http://www.netperceptions.com/) uses "collaborative filtering"--rules derived from the behavior of people with similar preferences--to determine what each visitor sees. (While Firefly is also known for its collaborative filtering technology, it announced plans to discontinue it after the Microsoft acquisition.) Both neural networks and collaborative filtering offer the promise of rules that can evolve automatically with changes in user preferences, conditions and offers, overcoming the limitations of manual rule definition.

The final element in interactive relationship systems is the responses themselves. Most systems rely on the marketer to set up a database of "content", which is typically a series of prebuilt Web pages that will be sent to visitors as the rules determine. Some, including Aptex, Rubric and MarketFirst, can also generate other types of messages such as e-mail, faxes or even physical mail. Aptex is particularly notable for its ability to "understand" incoming text messages and offer appropriate replies, while MarketFirst has an unusual ability to mix several different media (e.g. fax and Web) in the same document. This type of cross-media response is another critical challenge that must be met as customers interact with companies through an increasing variety of channels.

Perhaps the ultimate in interactive relationship management is to provide custom Web pages that are tailored to the individual visitor's needs. BroadVision (650-261-5100; http://www.broadvision.com/), Micromass intelliweb (919-851-3182; http://www.micromass.com/) and Persimmon IT do this by assembling pages from templates whose contents are selected by system rules. This means that each visitor effectively sees a personal Web site, which is adjusted constantly as their interests evolve. The hope is this will prove so valuable that the visitor will keep returning, generating new opportunities to generate advertising revenue and make appropriate marketing offers.

No matter how advanced, even a personal Web site is just one element in an interactive marketing relationship. Today, no single product combines the most

Interactive Relationship Marketing

sophisticated methods for handling vistor actions, marketing rules and responses. Although some cooperation is possible among certain systems, most operate in isolation--due primarily to their separate internal databases. We can expect this to change as the industry matures and vendors realize that a single product cannot hope to lead in all these different areas.

Chapter 3: Optimizing an Online Campaign

How Internet Advertising Works

Title: How Internet Advertising Works
Author: Rex Briggs, Vice President, Millward Brown International, San Francisco, USA and Horst Stipp, Director, Social and Development Research, NBC Television Network, New York, USA
Abstract: This presentation reviews the research on the impact of different forms of online advertising, presents the authors' analysis of the findings, and suggests directions for future research. The main conclusions are:
(1) advertising on the Internet can be very effective;
(2) for most forms of advertising, the Web should not be regarded as a substitute for advertising on traditional media. Rather, indications are that Internet advertising works best if it is part of a coordinated campaign that includes traditional media.

Copyright: © Millward Brown International

Introduction

The rapid increase in the number of people who can and do access the Internet, especially in the US and in Europe, has resulted in similarly strong increases in the amount of money spent on online advertising. This increase in spending makes it more and more important that the effectiveness of these new forms of advertising is explored and that we understand how they work.

The body of research that has explored if and how Internet advertising works is still comparatively small, but it is growing rapidly. We will review this research, including up-to-date unpublished studies, analyze its conclusions, and suggest directions for future research. However, before we do that, it is appropriate to define what we consider "Internet advertising."

Internet Advertising

We distinguish between three broad kinds of advertising on the Web; they are:
1) placed ads (including banners, Rich Media ads, pop-ups, animated cursors);
2) sponsored elements within sites; and
3) company marketing sites. There is, of course, other online marketing activity on the Web. Company sales sites (similar to product catalogs with order forms) are one example. However, in our opinion, sales sites and sales banners should be considered in a direct marketing context rather than in an advertising context.

Research on the Effects of Internet Advertising

During the last three years, a large number of research projects have been conducted to explore the effectiveness of the different kinds of online advertising. We will review four kinds of studies in this field:
1) research on the impact of banners,
2) research assessing the effects of

clickthrough (which includes studies on entire Web sites),
3) research comparing the effect of different executions (such as which kinds of banners work best), and
4) studies comparing the impact of Internet advertising with that of ads in other media.

1) The impact of banners.
The first study conducted on the subject was done on behalf of HotWired by Millward Brown Interactive in the fall of 1996 (Journal of Advertising Research). Prior to this study, it was widely held that banners were simply direct marketing enticements. This study demonstrated that advertising communication occurred even when users did not directly respond by clicking on the ad. The seminal study has been replicated a number of times. (See 1997 Electronic Telegraph/Ogilvy & Mather, or 1997 Internet Advertising Bureau (IAB) Online Advertising Effectiveness Study, for example). The same experimental design used in these studies has been extended to measure Interstials (Berkeley Systems' You Don't Know Jack, the Netshow, 1997), sponsorships (Dockers Dream Jobs 1997, NBC 1997, Synapse Oldsmobile 1998, Intel 1998), Rich Media (InterVu 1998, Wired Digital 1998), and even the cursor space (Comet Cursor 1999). In each of these carefully controlled studies it has been demonstrated that online advertising can boost brand awareness, positive brand perceptions, and intent to purchase.

2) The effect of clickthrough.
The first study of clickthrough simply measured what wear-out patterns emerged due to time and frequency (DoubleClick, 1996). The Internet Advertising Bureau study, conducted by Millward Brown Interactive (1997,op.cit), delved deeper and measured attitudes toward a variety of brands among those Web users who were observed to click on an ad and those who did not. The research revealed that clickthrough was usually motivated by consumer believing that an immediate need should be satisfied at the moment they saw the ad by clicking on the banner. Because many categories do not have an impulse need, or cannot fill such a need online, clickthrough is not an effective predictor of advertising effect. This, apart from lack of interest in a product, appears to be an important reason why clickthrough is relatively rare. (AdMap 1998)

3) Does creative execution make a difference?
The first study on this subject was presented at the second Internet Advertising Summit in 1996. It was conducted for J Walter Thompson Specialist Communications by Millward Brown Interactive and it measured the effects of different creative execution of Job Opening ads. Since then, a number of well-designed studies have been done on this topic. For example, the IAB study (1997, op.cit.) reports an analysis by Scott McDonald on "What seems to be working." The 1997 Electronic Telegraph/Ogilvy & Mather study compared four creative executions for two products and found different effects. Below are the two ads for IBM that were tested.

The two ads produced statistically significant differences in performance.

Figure 1: ad A

Figure 2: ad B

Figure 3

The chart on the right shows the difference in brand-linked recall of the two IBM ads versus the control ad.

The results demonstrate that creative differences had an effect above and beyond differences in the media context and target audience, though these variables are also critically important to the success of the ad.

A more recent study conducted at Michigan State analyzed ad position on a page. The 1998 InterVu study (1998, op.cit) observed the BrandImpact of two nearly identical ad executions on the CBS Sportsline Web site, but one had the addition of a 1 x 1 inch streaming video of a TV ad. See figure 4.

4) Internet vs. other media.
We found only two published studies that focus on this topic (Dreze and Zufryden, 1998; Bezjian-Avery et al., 1998). These authors emphasize the limitations of the Web as an advertising medium. However, they do not consider aspects that we will discuss in connection with other research later in this paper.

Research Review: Internet Advertising Works

As in all advertising, one should avoid simplistic general conclusions in the area of online advertising. Target audience, creative execution, the specific purposes of a campaign, and the situation and needs of the brand are factors that make every advertisement and its impact unique cases. Having said that, we do think the research evidence suggests a number of general conclusions:

- The most important finding is, unquestionably, that Internet advertising works. The research has shown that Internet advertising can increase sales, enhance the image of the brand, affect the consumers' attitudes towards the product, and it can impact new information about products and brands. Needless to say, not all advertising in this new medium is effective - just as not all advertising in traditional media achieves the desired outcome
- Further, the research establishes that clickthrough is not necessary for effects. In fact, since clickthrough is relatively rare, only a small portion of Web advertising effects can be attributed to

Figure 4

How Internet Advertising Works

clickthrough. The most important Internet advertising tool is the banner.

- Reaching the right audience is key. Research suggests that online users are actively browsing for information. If the ad is not perceived as relevant to the consumer, he or she will pay little attention and the effect will be minimized. This is similar to print advertising, but in contrast to some TV advertising, which can be quite effective if the consumer perceives the ad to be entertaining.
- Research on different executions indicates that, as in other media, creative execution plays a major role and can impact the effectiveness of a banner significantly. Finally, the one published study that compares Web advertising with ads in other media focuses on interactivity, that is, clickthrough rather that simply exposure to banners. The authors of the study conclude that interactivity may not always be advantageous for advertising. However, as said, the other research demonstrates that interactivity is a relatively small aspect of Web advertising. Clearly, more research is needed on the subject.

New Directions in Research on Internet Advertising

One of the most remarkable aspects of the research on Internet advertising is the fact that practically all studies deal with the Internet in isolation, other media are ignored. In fact, the studies often control for outside forces through experimental design to enhance the accuracy of the measurement of the value of Online advertising. Given that this is a new medium that needs to establish itself as an advertising medium, this is quite understandable. But that does not mean research should continue in this manner. In fact, analysis of the data of a study on Ford banners in the 1997 Electronic Telegraph study sounded the wake up call for Millward Brown Interactive. In this study the BrandImpact of the following two ads for Ford were observed:

Figure 5

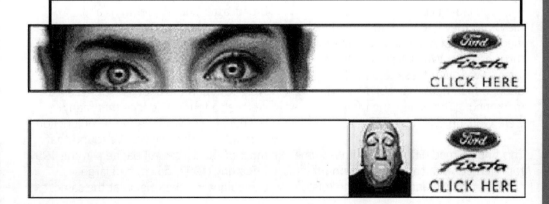

The BrandImpact of these ads were observed against a control among users accessing the Electronic Telegraph Web site. The ads worked - they increased consumers' perception that Ford is a safe car. This baffled the American analyst: How could this banner ad, which contained nothing but an odd picture and the Ford logo convince anyone that Ford is a safe car. The research design was sound: Those users who happened to access the Electronic Telegraph during the course of the study were randomly sampled, assigned to control or test cells and delivered the ad (without any alteration to the Web site). After the user was exposed, an online survey asked them to answer a few questions. The results were clear - the ad boosted the perception that Ford is a safe car. Why?

What the American analyst from Millward Brown Interactive did not know was that Ford was running a heavy TV, print, and outdoor campaign that used the same creative found in the online banners. Those other ads contained additional messages about safety features including halogen head lamps (the first ad shown above) and air bags (the second ad). The online ads were reminding consumers of the messages they had seen and heard in other media. Not just reminding, but reinforcing the messages. When Millward Brown Interactive found these results, they dubbed the effect "surround-sound marketing." The online and offline ads were working synergistically.

This finding made it clear that it was time for Internet research to look beyond the computer screen and consider the role of the Web in conjunction with the other media.

The most compelling reason for broadening the scope of Internet advertising research is the usage of the medium by the consumer. Practically all users of the Internet are users of other media. Through Millward Brown Interactive's VOYAGER Web Panel, a nationally-representative panel of US Web users drawn purely through Random Digit Dial, we found that 90% report watching TV "yesterday," 88% listened to the radio "yesterday," 77% subscribe to a magazine, and 58% subscribe to a newspaper. In fact, most Web users report watching TV more hours than they use the Web.

Television networks have been aware of the fact that TV and Web usage are very compatible for some time and they have used television to build traffic to their Web sites. This was first documented in a paper by Coffey and Stipp (Journal of Advertising Research) and it is evident from the large amount of traffic to sites such as CNN.COM and MSNBC.COM. Proprietary research by NBC shows that network Web sites can enhance the image of the broadcaster and its programs and strengthen the fans' commitment to the shows and its sponsors.

Further, an analysis of the banners that were tested in the above-referenced IAB study also found evidence that many banners might work better in conjunction with ads in other media. We found that most of the successful banners in the 1997 IAB study (1997, op.cit) had large campaigns in other media at the same time. These were banners for Delta Business Class, Volvo, and Schick razors. One big media spender, Mastercard, did not do as well. Delta and Volvo appeared to use consistent imagery across media, -- It is not apparent whether Mastercard and Schick campaigns were coordinated across media.)

Are these findings sufficient to conclude that coordinate multi-media campaigns leveraging consistent imagery and include the Internet are likely to be more effective than Internet-only campaigns? Or that campaigns in traditional media can work better if Web banners are added? Not quite. Though it is a good hypothesis. We need more research in this field, especially to identify the kinds of products and the circumstances under which multi-media campaigns will work better. On the other hand, we don't think we are sticking our necks out too far: There is thirty-year old research (which was corroborated about five years ago) which provides empirical as well as theoretical support of our hypothesis (references). The concept is called "imagery transfer" and it is based on studies that found, for example, that radio ads using a TV soundtrack evoked images of the commercials among listeners who had seen the TV commercial. We are suggesting that similar processes can work (and are working) online.

Summary

We believe that the findings from the first studies on the impact of Internet advertising in the context of multi-media campaign interactions should be high on the agenda for future research efforts in this field. Since we assume that the Internet will not replace traditional media (like television and radio), it means that there will be advertising in a variety of media. How the growing new medium, the Internet, fits into the media mix, is a most urgent issue for all advertisers. If it is true that many products and companies could benefit from coordinating their TV, print, and Internet advertising, this research would provide extremely valuable information that would have a positive impact not only advertisers, but also on the health of the new and the traditional media.

Is Your Online Marketing Working?

Are you reaching the right audience? Do they remember your advertising? Does your marketing enhance and sustain positive associations with your brand? Does your marketing drive sales? In sum, are you getting a positive return on investment from your online marketing? We can answer these questions for you. This paper explores how Millward Brown's strategic approach to measuring the impact of your marketing activities, specifically your online marketing activities, can provide a key competitive advantage. With our Voyager service, we work as an integrated part of your marketing team to enhance your strategy and tactics.

Impact Marketing

In today's sophisticated marketing world, any advertised category is in a state of dynamic change. Companies battle constantly for sales and midshare. Getting feedback on the state of the brand - from the consumer's point of view - is critical to answering the key questions of marketing effectiveness and return on investment. Beyond listening to consumer feedback, Millward Brown Interactive works with you to gain knowledge of how your online consumers think, feel, and behave in your category. Most importantly, Millward Brown Interactive relates your marketing activities (designed to influence consumers) to changes in consumer attitudes and purchasing actions. This insight will help you understand what works and why. Drawing on decades of experience across a range of product and service categories, Millward Brown is able to contextualize your brand's experience in the marketplace. We call this combination of research methodologies and expert analysts our Voyager service. This paper addresses the WebATP (Web Advanced Tracking Program), Voyager's cornerstone.

We endeavor, through the WebATP, to clarify for our clients how well each of their marketing activities is working (and has worked). Over time, the result is an informed and effective contribution to the planning and direction of our clients' future advertising and marketing strategy and tactics. The partnerships Millward Brown Interactive builds with clients - based on gathering consumer feedback, recording action, and interpreting results - provide tremendous insights and positive return on investment.

Insight, Answers, and Ideas

Millward Brown's approach and process are time-tested and proven. Two decades ago, we invented the advanced tracking approach to measure the impact of traditional advertising. Now Millward Brown Interactive has extended this proven approach to online marketing. At the highest level, Millward Brown Interactive provides a concise map relating your marketing activity to its impact on the consumer. This map (a graphical output) relates trends in the consumer attitudes and behavior directly to the exact timing and weight of your marketing activity. For example, in the following chart, the marketing activity is charted on the bottom half with the corresponding consumer response on the top half. This makes the impact of marketing activity readily apparent. See how consumer brand awareness trends upward over time in relation to television advertising (TV GRPs)?

In general, this type of insight allows the client to judge which element of the marketing program affected the observed trends in consumer response. In this example (drawn from Millward Brown International's Advanced Tracking Program of traditional advertising) it is clear that TV advertising was responsible for raising awareness of the brand. (Note the immediate rise in ad awareness and brand awareness coinciding with the first flight of TV advertising.) If you look at the "claimed trial" trend, you'll notice that it is largely unaffected at the start of TV advertising - it begins to trend upward with the introduction of promotions. In other words, promotions generated most of the trial. This insight is extremely important because it tells the marketer how to coordinate marketing elements to generate awareness and trial to enhance the success of a new product introduction.

Tracking online marketing activity is every bit as important as tracking traditional marketing mix variables. The added advantage of online marketing tracking is the ability to link specific online advertising exposure to a specific consumer response. Millward Brown Interactive carefully tracks the online activity of a large subset of the online consumer population. Compared with traditional marketing, online marketing affords even more precision in the measurement of advertising exposure and its results.

Think of the marketing decisions that confront you and your brands today. Do you struggle with the issue of how to communicate your positioning? Do you wonder if your consumers are absorbing your message? Do you worry whether or not your brand name is the one consumers think of first when they think of buying a product in your category? Do you hear the patter of competitors' footsteps quickly approaching your advantaged position in the mindspace of your consumers? These are the questions that keep good marketers up late at night. They are also the questions Millward Brown Interactive can help you answer. We've developed key measures for advertising awareness, crucial brand associations, competitive positioning, and sales to address critical questions of marketing effectiveness. Because we have tracked these metrics across a broad range of clients in a variety of categories, we can identify patterns in the marketplace and provide valuable insight into consumer reactions to marketing tactics. For example, patterns in the Advertising Awareness Index clearly indicate whether the creative execution of your advertising messages is strong enough to achieve your objectives (more on the Advertising Awareness Index later).

Making Sense of Advertising Awareness

Over the years, Millward Brown has built up the largest database of systematic continuous tracking in the world. This database indicates - beyond all reasonable doubt - that advertising must satisfy two basic criteria in order to be effective: it has to be both memorable and branded. It has to be memorable in order to catch the consumer's attention in a cluttered advertising environment and it has to be branded in order to ensure that people remember the advertisement in relation to the specific brand sponsoring it. Our experience measuring a wide array of online advertising reinforces this point. You want to know if your marketing is reaching the right audience. You want to know if your audience remembers the advertisement. You want to know if the ad enhanced audience perception of the brand. And you want to know if the audience bought your product. We provide the answers to these critical questions. Better yet, we provide ideas about how to improve strategies and tactics - acting as a sounding board and a partner in the quest to enhance your return on marketing investment.

Underlying our approach is one fundamental principle: to understand consumer response, we must listen intently - and continuously - to the consumers. This means we use a structured survey that we field every day of the year because periodic waves of surveys can miss important shifts in the marketplace. We identify short - and long-term trends and relate marketing activity to changes in the marketplace, monitoring both your brand and the brands of your competitors. Through the WebATP, we deliver critical insight and competitive advantage, enabling better marketing decisions. We monitor not only what consumers say, but also what they have been exposed to online and what they do in the marketplace.

In Continuous Touch with Your Consumers

In today's rapidly changing marketplace, if you blink you're going to miss something. Online media has accelerated the pace and made constant contact with consumers even more important. Traditional pre- and post-wave studies can mean your eyes are closed to the marketplace for prolonged periods of time. The following chart demonstrates that a wave study carried out at points "A" and "B" would have missed the relationship between the key trends and different marketing activities - hiding the different contributions of TV advertising and consumer promotion. In essence, the marketer would not have learned that advertisers alone will not generate trial in this circumstance and that promotion is necessary. Are there any marketplace shifts that you are missing?

Short- vs. Long-Term Trends

By changing the time period across which the data is rolled, we can highlight short- or long-term trends in the data. We typically start with four-week rolling data to highlight short-term changes. As we can see from the graph above, (which shows unaided brand awareness for a well-known brand of candy) four week rolling data can be variable. We have both the varying effects of different marketing variables (such as pricing, trade, and consumer promotion) and normal sampling error at play. Often, we can gain additional insight by examining longer-term trends. When we roll the data across 16 weeks, a much clearer pattern emerges as longer-term trends are revealed.

Monitoring the Competition

A major advantage of continuous interviewing is that it enables you to monitor competitive activity on exactly the same basis as your own activity. Wave studies cannot do this because they are timed to monitor only your own activity and can take no account of the timing of competitors' actions.

At a basic level, the competitive data provides a context for judging the success of your own marketing activity by highlighting the difference between category and brand trends. Additionally, the information also allows you to evaluate the success or failure of your competitors' activities and to learn from their successes or failures before they do.

This has benefits at both tactical and strategic levels. It can provide early warning of competitor success and allow for effective countermeasures to be taken. In other cases, it can highlight competitive vulnerability and allow you to exploit it. Overall, our WebATP adds to your understanding of marketing activity that can be used successfully within your category.

Putting It All Together

It takes more than good data, however, to turn the WebATP into a truly actionable planning tool. Analysis, interpretation, management judgment, and experience are also key ingredients. To this end, management presentations and consultation are a critical part of the services we deliver. In these presentations, senior Millward Brown executives offer an informed interpretation of all the data from the WebATP. Data from the marketing questions are interpreted and definitive conclusions are reached on how the advertising and marketing are likely to influence brand awareness, perceptions, and purchase. The litmus test is to relate your particular marketing activity back to your marketing objectives. Specifically, does the advertising actually improve awareness or images? Does it change the status quo in the way that was intended? Does it yield a positive return on investment?

Answers

Because we strive to help you develop workable solutions to your marketing conundrums, an ongoing application of learning to improve your business is a central element of the service we provide. Each client has secure, password-protected access to key learning through our Marketing Action Web Pages (MAP). On these pages you will find a clear articulation of the marketing objectives and the research approach to quantifying success at achieving your goals. In addition, access to surveys, data, and management reports is at your fingertips 24 hours a day, seven days a week. We focus on understanding your business and ensuring that we are applying our knowledge, insight, and learning to work for you to build a smarter organization with better, more effective marketing. The litmus test for our effectiveness is your success.

An annual WebATP has many business applications. It can be used to:
- Assess your brand awareness relative to key competitors

Are you trending up? Down? If down, would a return to advertising or a switch to a more effective advertising be warranted? How effectively does each impression generate brand awareness? How well does awareness convert to trial?
- Determine and monitor your brand's positioning

What are the hidden strengths and weaknesses compared with the competition? A hidden strength can be highlighted in future advertising strategy. A hidden weakness can be addressed via reformulation, new product development, or countered with ad strategy. Does campaign effectiveness differ by targetable segments? We can help you focus the right messages to the right target audiences
- Evaluate your brand's equity

How is your brand perceived by your consumers? We can identify what steps should be taken to improve your place in the market.
- Examine purchase trends and product satisfaction

Are they trending up? This could be the result of new advertising copy, a change in weight or media flighting, or a successful promotion. We can help you determine what contribution each of these elements is making. Are they trending down? This could signal competitive pressures, pricing concerns, or a product quality issue. We can help you pinpoint the problem.
- Assess the effectiveness of your current advertising

Is it generating visibility for your brand? If it is strong and communicates well, we can help you make it work harder with better targeting. If it is weak, we can diagnose a more effective solution to replace it or improve its performance.
- Learn if the specific product attribute claim made in your advertising was remembered

Is it likely to benefit the brand in terms of changed purchase intent and attitudes?
HOW IT WORKS:
The mechanics of measuring the impact of your marketing activity involves:
1) Drawing a true cross-section of Web users
2) Continuously gathering consumer response to your brand and key competitor brands
3) Analyzing and modeling data to generate usable marketing insight
4) Linking the marketplace learning to your business objectives and strategy and making actionable recommendations

A True Cross-section of Web Users

In the first year of the development of the Voyager service, Millward Brown has interviewed over 1.6 million US consumers and asked them if they currently use the Web. These interviews were conducted on the phone based on a Random Digit Dial (RDD) methodology. Without question, this is the most extensive survey of Web users ever. As of Spring 1998, Millward Brown has called into 1 out of every 60 households and asked 1 out of every 100 adult consumers if they use the Web. The combination of one of the most representative sampling approaches with the incomparable scope of our surveying guarantees the most projectable and accurate portrait of the Web consumer - in both home and work environments.

Continuous Data Collection

By interviewing continuously, we can record the dynamics of a situation, highlighting exactly when things changed, and how. We interview enough people every day to balance your need for immediate short-term marketing feedback with reasonable time frames that enable marketer response while managing the cost to you. Typically, this means that we interview enough people to examine the data on a rolling four-week window and provide monthly data insight updates and in-depth quarterly analysis.

Rolling data works by aggregating data on a rolling basis. This approach is also known as moving averages. We aggregate data from weeks 1 through 4 and graph them with weeks 2 through 5, 3 through 6, and so on. This not only gives a high level of confidence in each point on the graph, but also allows us to identify short-term changes when and as they happen.

Finally, continuous interviewing allows us to bring the reporting of awareness and consumer attitudes toward your brand and examine them alongside common business statistics (e.g., sales and market share reports). This provides a more complete marketing dynamics portrait, which enhances decision making.

Analysis That Yields Insight

Perhaps the most important dimension of the Voyager service in general (and the WebATP specifically) is the focus on making sense out of the data so that you can make informed marketing decisions. This begins with understanding your brand and its category and it includes a focused understanding of how to measure and compare advertising and marketing effectiveness. It involves diagnosing what isn't working and recommending what to do about it. Voyager transforms data into insight that provides ongoing learning for your marketing team.

Understanding Your Category

Although there are basic guidelines (in terms of methodology and questionnaire design) that have evolved from over 20 years of real-life tracking experience, the design of each Voyager WebATP is customized to suit the category and brand to be tracked. The primary objective of each WebATP is to provide you with a real understanding of how people view your brand, and how its position is changing across time in relation to your marketing efforts. To this end, each WebATP includes a comprehensive set of questions designed to monitor and evaluate brand performance, including measures of awareness, usage, and brand attitude.

These sophisticated analyses have been developed to measure brand health and to show strengths and weaknesses in the brand's image profile. Over time, results from these analyses can be related back to your marketing activity to show which has successfully impacted your brand's standing in the market.

From Advertising Awareness to Advertising Associations

Our considerable experience with advertising tracking suggests that there is no one thing that can be defined as advertising awareness. Different questions access advertising awareness in different ways and can give very different responses. Millward Brown has spent considerable time finding the most sensitive and relevant way to access advertising awareness.

The question we ask is "Have you seen (brand) advertised on (media) recently?" The most important point about this question is that we access the advertising awareness via the brand name. This is crucial to our success in linking changes in advertising awareness to changes in brand attitudes. In the final analysis, what we really want to know is what images and feelings the advertising has managed to establish for the brand. By getting people to think about the brand in the context of advertising, we can access those branded advertising associations.

The question is much more than a read on whether or not your advertising is noticed, it is an important indication of your success at locking brand association in the minds of your consumers.

When we ask people whether they have seen a brand advertised in particular media recently, we get responses which move in consistent patterns across time, and which clearly relate to both the weight and quality of the advertising. Similar data is gathered for all relevant media, like television, print, and radio.

In the example below, we can clearly see that TV commercial "A" generated relatively little advertising awareness, while commercial "B" generates far more. The first two flights of each campaign are almost identical in weight, just over 700 GRPs (Gross Ratings Points - 1 GRP equals 1,000 impressions), but commercial "A" hardly changes awareness, commercial "B" pushes it up to over 30%.

Not all campaigns provide us with such clear-cut findings. Recall of past advertising and differing weights and patterns of spending make it more difficult to assess whether one commercial is leveraging advertising impressions into branded advertising awareness better than another. It is for this reason that Millward Brown has developed a means of modeling advertising awareness against advertising impressions (normalized into GRP expenditure).

Modeling Advertising Awareness

Based on tracking more than 10,000 advertisements, Millward Brown has developed an analytic model that quantifies how well different commercials generate branded advertising awareness.

How Internet Advertising Works

This is not as easy as it sounds. At any one time, the level of advertising awareness is a function of three things:
- Historical advertising
- Media weight
- Quality of the advertising message

These three elements must be disentangled to arrive at conclusions about current advertising effectiveness.

The basic aim of the modeling is to relate incremental increases in advertising awareness to advertising expenditure (normalized into GRPs). The key statistic generated by the modeling is the Awareness Index. This tells you how much incremental branded advertising awareness is generated by 100 GRPs for a particular advertisement or campaign. This means that clients can compare the effectiveness of different campaigns to determine how well each works at communicating a branded message.

The other measure generated by the modeling is called the Base Level. This is the long-term component of advertising awareness. It is the level to which advertising awareness will fall in the short term in the absence of any current advertising expenditure. The Base Level tends to rise during memorable advertising, or periods of high advertising spending. (On the other hand, Base Level tends to fall in the longer term if current copy is not memorable or advertising expenditure is low.)

If we refer to the example of advertising awareness that we presented previously, we can now quantify the difference between the two ads. The modeling shows that Ad "B" generated 12 times the advertising awareness of Ad "A" from the same number of GRPs (Awareness Index of 6 for ad "B" compared to 0.5 for Ad "A").

The basic conclusion drawn from the modeling is that the ability of commercials to leverage GRP spending can differ much more than copy testing would lead you to believe. In real life some executions can actually deliver up to 20 times the advertising awareness that others do.

In the Awareness Index we have a real measure of the power of an ad to generate advertising memories that are closely linked with your brand. This is more than just a good advertising anecdote. In practical terms, this means that some ads work 20 times harder for the brand. There is a competitive advantage to knowing which ads are a hit and which fall flat - in the end you can get more for less if you have this knowledge.

Given the GRP data we can model Web, TV, and radio advertising awareness. We have also developed the means to model print advertising awareness (taking into account first-time readership accumulation and re-readership). This means marketers can understand the contribution of each element of the media mix as well as the synergy across media and advertisements.

Assessing the Quality of Advertising Associations

Beyond the quantity of advertising impressions, it is the quality of those associations that really matters.
In order to understand the nature of the

associations generated by advertising, we follow up the awareness question by asking people to describe the last advertisement they saw for the particular brand that they claim to have seen advertised. After this we ask them to tell us what impressions the advertising gave them about the brand. The purpose of getting people to describe the ad is not because we believe people mentally "scroll" through the advertisement at the point of purchase. We do believe, however, that the associations gained from advertising have an influence, and we use this questioning to get a clear picture of those associations. The recall part of the question not only focuses people's minds on the advertising they have seen, but also serves a useful diagnostic purpose - it allows us to see which elements of an ad are the most memorable. If an ad fails to meet its communicated objective, or if it is not well-branded, it is usually because the message or brand name was not integrated with what was memorable. Advertising stands the best chance of influencing brand attitudes and brand choice if the associations it is designed to put across are firmly linked to the brand at the time of seeing the ad.

Our theory of advertising effectiveness is this: you have to spend something to achieve anything, but the effectiveness of what you spend relies on:

1) The ad's ability to generate branded advertising awareness
2) The ad's ability to make a relevant claim or impression
3) The ad's ability to do so in an appealing or motivating fashion

Extreme care is taken to ensure that we gain an accurate understanding of consumer impressions. This enables us to produce insight based on those who can describe the current campaign and look very closely at how they describe it and what messages they got from it. Consequently, we can check communication much more precisely than one can from data that mixes up people who are actually describing several different campaigns. Proven recall, therefore, is used to evaluate communication, not to evaluate awareness.

Consumer Promotion

In the current marketing environment it would be unrealistic to neglect the role of promotions. For some brands, promotion is an increasingly important part of the Web marketing mix. At Millward Brown we have routinely included the timing of consumer promotions to the marketing expenditure portion of our charts. We also recommend including questions on promotion awareness in the interview, since these have proven useful in explaining changes in brand-related data that might otherwise be attributed to advertising or another cause.

Achieve Your Business Objectives Through Marketplace Insight

No other company can offer the same accurate measurement of Web consumers, the level of experience in the analysis of continuous tracking of marketing data, or the same ability to turn that data into real findings for your business. Through use of techniques such as the modeling of advertising awareness and other proprietary analyses, we really can turn numbers into insight and consult with you as a partner in achieving your business objectives.

While no other company has done more to measure online advertising and online marketing effectiveness, it is our insistence on the close personal involvement of senior executives at all stages of the research that matters most. The benefit such commitment and expertise brings to our clients is concise and clear counsel because we understand the marketing process online and offline and we understand your specific business objectives. Our senior executives can cite precedents and draw parallels from our in-depth knowledge of other research findings to provide you with relevant feedback that enables good, properly informed decisions to be made.

In conclusion, the WebATP provides the best available assessment of how marketing activities benefit your brand and company. It is a proven technique that can contribute at a fundamental level both to your tactical and strategic decision making.

Given the value of the Advanced Tracking Program (ATP) Millward Brown pioneered, it is little wonder that over one-third of the top 100 advertisers in the US rely on Millward Brown to measure return on investment from their marketing activities. In sum, we can benefit your brand with our insight by using those measurements to help you improve return from your online marketing investment.
To learn more about what the WebATP can do for you, please contact Mark Thomas, Millward Brown Interactive's vice president for sales. mark.thomas@mbinteractive.com.

A ROADMAP TO ONLINE MARKETING STRATEGY

"Accountable media" has become the battle cry of advertisers. Three years ago, when the first Web banners were sold on HotWired, Rick Boyce - a media planner turned HotWired's cybersalesmen - observed, "We have the first truly accountable advertising medium! We can literally count each consumer who responds to the ad banner with a clickthrough to the advertiser's Web site." Many marketers were captivated by the banner's powerful ability to allow direct and instantaneous response to its offer. Clickthrough - a direct marketing manifestation - quickly became the standard for evaluating the effectiveness of Web ad banners. As Rick and I were first to discover when we conducted the 1996 HotWired Advertising Effectiveness Study, clickthrough may not account for the brand enhancement that results from exposure to ad banners. More than a year and several rigorous research studies later, we can be much more definitive: CLICKTHROUGH IS THE WRONG METRIC TO ACCOUNT FOR BRAND ENHANCEMENT. The Web excels at direct marketing, but it does much, much more. Even as advertisers come to grips with what clickthrough does and does not measure, brand managers struggle to redefine online marketing strategy. This paper will encourage you to think about how online advertising works by putting clickthrough in proper context. From the strength of this perspective, we provide a roadmap to online marketing strategy.

What is the value of clickthrough?

The clickthrough metric (the percentage of those exposed to the ad banner who click on the banner to connect to the advertiser's Web site) is important in measuring how banners work in terms of direct marketing. However, when this metric is used to measure the effectiveness of advertising communication, clickthrough fails grotesquely. Why is this so? After all, many have argued that if someone is truly affected by an advertisement, he or she will click on the banner and go directly to the advertiser's Web site for more advertising information. The point of view that Web ad banners are really little "ads" for a bigger ad that is only delivered when users clickthrough is not completely unreasonable. It is generally true that a user who takes the time to transfer over to the advertiser's Web site will take away a heightened level of information regarding the advertiser. However, there is a flaw in the logic that Web ad banners have little value that stems from a handful of incorrect assumptions, namely:

- Brand enhancement can only happen on the advertiser's Web site
- Brands have far more to say than could ever be conveyed in a Web ad banner
- Clickthrough should be sought by all advertisers
- The level of clickthrough determines the level of brand enhancement

These assumptions are both incorrect and counterproductive, as we will demonstrate. For the brand manager to make the most of Web investment, we must put clickthrough in its proper context and

How Internet Advertising Works

examine the development of a superior online marketing strategy.

Brand enhancement can happen as a result of exposure to an ad banner alone

While additional powerful messaging may await the consumer on the other side of an ad banner, the ad banner itself does a significant amount of brand enhancement communication. In fact, among the twelve ad banners we tested as part of the 1997 IAB Online Advertising Effectiveness Study the value of the ad exposure is significantly greater on average than the value of the clickthrough.

To put the clickthrough metric in proper context, consider that "recall of the advertisement" (in other words, ad awareness) was boosted by four-tenths of one percent (from 43.7% to 44.1%) as a result of those who clicked on the ad banners in the 1997 IAB Advertising Effectiveness Study. That's an increase of less than one percent! This means that 96% of the boost in Ad Awareness was caused by the ad exposure alone. The remaining four percent was caused by the clickthrough. This pattern is consistent across the various brand enhancement metrics we examined.

The model that assumes the user is passively exposed to Web pages and then, upon exposure to the right Web advertisement, miraculously shifts to an engaged mode upon clicking through is not supported by the data. Rather, the effectiveness of Web advertising seems to stem from the fact that Web usage is an actively engaging exercise, similar to reading magazines. Users are fairly attentive to the media environment - including the advertisements. Clearly, the belief that the ad banner is really a small

Figure 6

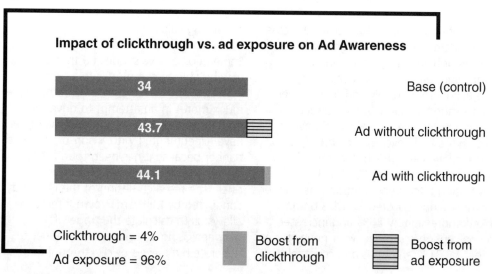

ad for the "real" ad that waits on the other side of a clickthrough can be rejected based on the analysis.

Many brands have a straightforward message that can be conveyed in a Web ad banner. Many advertisers have developed a clear and concise brand message or theme that consumers have heard any number of times: "Behind every healthy smile is a Crest kid," "Built Ford Tough," or "Amazon.com: Earth's biggest bookstore," for example. Being clear and concise is important if the advertiser is to lock a brand message into the consumer's long-term memory and have them recalled in the appropriate circumstance. In general, these messages can be communicated within a Web ad banner and do not require the consumer to transfer over to a Web site for additional elucidation.

These messages set up expectations for brands so that the consumer is more willing to try the product when in an experimentation mood (in the case of packaged goods) or is more likely to investigate the product further (in the case of considered purchase categories). If clickthrough works by presenting a compelling reason for the consumer to take action immediately and online, then what - short of changing its product - would a brand like Schick razors have to do to get a high level of clickthrough? Consider the case of Schick razors after they were tested in the 1997 IAB Advertising Effectiveness Study. The banner exposure alone boosted Schick's brand-linked impression by 26% and increased the perception of "meets your needs for a razor" by 31% and "has an acceptable

price" by 28% - these are all statistically significant increases at the 90% confidence level. Every level of the BrandDynamics Pyramid (a Millward Brown Interactive metric that measures the relationship consumers have with the brand) is enhanced, and the Consumer Loyalty Score (an accurate measure of the likelihood a consumer will purchase the product next) increased by 4%. Yet the clickthrough is only 0.5%. Does the low clickthrough rate imply that the Schick ad was a failure? Focusing solely on the clickthrough rate might lead the advertiser to conclude that this ad was not a success - which is contrary to the evidence from multiple brand enhancement and ad effectiveness measurements. Certainly, there are applications of direct marketing and promotion that can create and fulfill an immediate need online. This capability is one of the tremendous assets of online. In these circumstances, clickthrough has meaning; however, when it comes to communicating a straightforward branding message, clickthrough has little to no relevance. Exposure alone has brand-enhancing value.

Some people have suggested that a banner may be able to communicate a brand message, but a Web site can communicate much more! In an attempt to drive more eyeballs to the Web site, a few advertisers have experimented with a non-branded ad banner because it has been found that such ads generate higher clickthrough rates. The research design of the IAB study conducted by Millward Brown Interactive allows us to calculate the trade-off between using a non-branded banner to generate high levels of clickthrough and using exposure to generate brand enhancement. An advertiser who sacrifices the brand message on the exposure level in hope of achieving better brand enhancement by bringing more people to a dedicated Web site would need to achieve stratospheric 26.4% clickthrough rates to do as well as a branded ad banner with zero clickthrough to the dedicated Web site.

But what about using clickthrough as a surrogate measure (or predictor) of brand enhancement generated from the exposure? We certainly have heard of cases where an advertiser has said, "ad 'A' must be better than ad 'B' because it received higher clickthrough." BEWARE! We found a pathetic -.02 correlation between clickthrough and ad recall, suggesting that clickthrough does a very poor job of predicting the level of brand enhancement. Why does clickthrough fail to measure the brand enhancement value of an online ad banner? Our analysis indicates that consumers click on ad banners because the ad is not only relevant and engaging, but because they perceive an immediate need (stimulated by the banner) that can only be fulfilled by clicking through to a Web site where the promise made in the banner is delivered. We can think of many situations of effective advertising where the brand communication is relevant and engaging yet there is no credible reason to require a consumer to take immediate action - online is no exception to this principle. This principle (highlighted in the chart) explains why clickthrough does not predict brand enhancement. Clickthrough measures a direct marketing phenomenon, not a branding one. While the Web's direct marketing capabilities are exceptional, there are many brands for which this model does not apply. These brands should focus on the branding benefits of online.

Figure 7

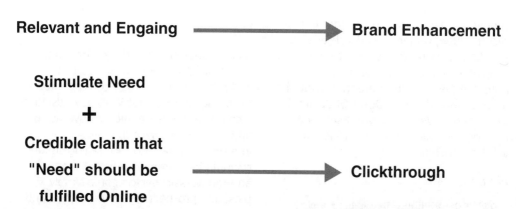

How Internet Advertising Works

Clickthrough should not be sought by all advertisers. There are some situations where clickthrough is directly relevant to marketing objectives. Few would argue against the idea that clickthrough is important for Web based products and e-commerce services fulfilled via the Web. Millward Brown Interactive does a significant amount of research for companies competing in this arena and advises our clients that clickthrough is often the beginning of a critical sales - and relationship-building process. For products that are not Web-bound, there is certainly a legitimate question as to whether or not clickthrough should be the goal of the advertisement. Consider this scenario: Millward Brown Interactive's tracking of Web behavior through our nationally representative panel has demonstrated that a significant amount of Web usage occurs at work during work hours. Consider the opportunity for a snack brand to communicate to hungry working digerati. Perhaps the ad might show a candy bar package and ask, "Is hunger making it hard to concentrate?" The advertiser should not necessarily encourage users to click through; rather, the advertiser's goal should be moving the worker to the vending machine where he or she can purchase the advertised brand. Maybe the ad copy should read, "Don't Click Here! Devouring a Web page won't satisfy your hunger, This will ... Now where is that vending machine?"

Figure 8

Undoubtedly, some people will click on the banner; however, the ad's goal is to reinforce top-of-mind consideration of the product and, perhaps, generate a behavioral response other than clickthrough (namely purchase). Given this objective, measuring the success of the campaign by the number of clicks would be inappropriate. In fact, a brand pursuing this strategy should ask what need are we accidentally tapping into that would cause someone to take an undesired action like clicking through!

What about high-consideration categories? What about high-consideration categories like automobiles? Should clickthrough be the critical metric when evaluating the success of an automobile Web banner campaign? The answer to this question is no. We believe the practice of evaluating automotive Web advertising on the basis of clickthrough could be compared to assessing television ads for automobiles on the basis of how many people visited the relevant showroom the next day. It is an ideal response, but not the most common one.

So how should a brand like Volvo think about clickthrough? We suggest that Volvo and other brands in high-consideration categories think about Web advertising as a combination of traditional advertising and face-to-face selling. In an interpersonal, face-to-face selling situation it would be extremely efficient if an automotive salesperson could simply present a prospect with the contract and ask the prospect to "buy right now." Efficient, but unlikely.

How Internet Advertising Works

Consumers generally require a dialogue that evolves from establishing the relevance of the product, to demonstrating performance, to discussing the advantage vis-a-vis competitors, before the consumer will become bonded with the brand. This evolution of the interpersonal selling dialogue is substantially similar to the hierarchy we use in our BrandDynamicsTM Pyramid. We believe that Web advertising can be used to drive consumers through this relationship enhancement cycle. It is only in the later stages of the cycle that it is important for the user to become directly engaged with the brand at the advertiser's Web site or, in the physical world, the dealer's showroom. The Volvo ad we studied demonstrates this point succinctly. The ad banner yielded less than one half of one percent clickthrough, yet the Volvo ad banner generated statistically significant increases on key metrics of product "relevance," "performance," and "advantage vis-a-vis competitors." Given the low clickthrough, was the Volvo ad banner a success? We think so.

Figure 9

The BrandDynamics™ Pyramid

Bonding
Advantage
Performance
Relevance
Presence

Clickthrough has its place - but it is the wrong metric for measuring brand enhancement. Indeed, there are many approaches to enhancing brands and driving sales among online users. Online marketing is much more than building and supporting a corporate Web site.

Roadmap to marketing online

While leading Web advertisers have come to grips with the new insight that the Web can work as both a direct marketing and a brand enhancement vehicle, many others are struggling to form a coherent Web marketing strategy. While the direct marketing model is a legitimate and effective use of the Web, it is Millward Brown Interactive's perspective that those who are leveraging the Web only as a direct marketing vehicle may be significantly under-utilizing their online presence and leaving themselves vulnerable to their competitors.

The proven capability to enhance brands with exposure to online advertising creates a new landscape of possibilities for marketers. But this landscape is fraught with both opportunities and threats: opportunities to use exposure to online advertising messages to make consumers aware of your product, to change consumer perception of your brand, and to take market share from your competition; and threats that your competitors will figure it out first - or be more effective in their execution of online advertising. How can you successfully navigate this new landscape? The brand manager is faced with two critical questions: Should my brand include online in the marketing mix? And if so, how?

What do brand managers really want to know when deciding to invest online? They

How Internet Advertising Works

want - and need - to know that the people they are reaching have economic value and that reaching them online is an effective use of their marketing dollars.

Should your brand include online in your marketing mix?

From my perspective, answering this question is straightforward: Measure the economic value to your category that online users represent. Take the example of an automotive manufacturer: 23% of the manufacturer's target market is online and, because of the Web's somewhat affluent demographics, these Web users are more likely to purchase higher-margin luxury automobiles - and therefore account for 40% of the company's profits (economic value). Garth Hallberg from Ogilvy & Mather suggests in his book All Consumers Are Not Created Equal that a marketer should spend marketing dollars in proportion to the segment's profitability. In this case, that means the brand manager should spend 40% of their budget reaching these online consumers.

So should brand managers spend the full 40% of their budget online? The answer is no. After all, online consumers are accessing other media as well. But spending all your marketing budget offline to reach online users as they consume other media isn't wise either. We know from extensive research that each media can work in a unique manner to enhance the brand. Indeed, advertising across media can have synergistic results, as we observed in the Electronic Telegraph/Ogilvy & Mather Online Advertising Effectiveness Study . We

suggest that brand managers consider "Surround Sound Marketing." Find the ways in which you can leverage each medium your consumers use to present and enhance your brand.

If brand communication in each media had an equal impact on your bottom line, we would suggest examining the share of media time these users spend online. To continue the example, if the automobile manufacturer's Web-enabled consumers represent 40% of profits and spend 15% of their media time online then it would be reasonable to earmark 6% (40% X 15% = 6%) of total media dollars for reaching and communicating with these consumers online.

Unfortunately, each media does not make an equal contribution to your brand and there are many variables at play that cause dangerous blanket generalizations regarding the relative effectiveness and the proper mix of media. Brand managers should determine budget allocation using a combination of the "share of profit/share of media" time analysis and solid research measuring the relative effectiveness and synergy of each media in building their specific brand. Finding the right balance will take experimentation and focused research. There is grave danger in complacency. If online advertising will have an impact and you choose not to exploit it, but your competitor does, you may be sacrificing your brand equity in the long-run and jeopardizing sales in the short-run. As many online marketers have experienced, there are many approaches to marketing online. Some of these strategies are well worth your hard-earned marketing dollars, while others may be a waste of resources.

How Internet Advertising Works

Which online strategies are right for your brand?

It is constructive for marketers to think of the Internet as nothing more than a technology that enables information transfer. Because of the unique characteristics of the technology, communication can be fashioned along two key dimensions: The first dimension is a continuum that ranges from proactive to reactive communication. The second dimension ranges from broadcast communication to personal dialogue. The chart below illustrates the dimensions and provides examples associated with communication types.

What does this way of thinking imply for the marketer? It means that you can reach and communicate with your online target in a multitude of ways - from a broadcast advertisement that proactively reaches your target audience on a mass-reach Web site to a personalized email delivered in reaction to an individual customer's query. It means that online is not monolithic. It is multifaceted and requires an online communication mix appropriate to your brand's marketing strategy. The effectiveness of your online strategy depends upon achieving the right mix for your brand. But with many options and an evolving technology that seems to expand the list of options on a daily basis, how is a marketer to develop a coherent strategy and effective tactics?

Development of a coherent strategy and effective tactics requires the marketer to

Figure 10

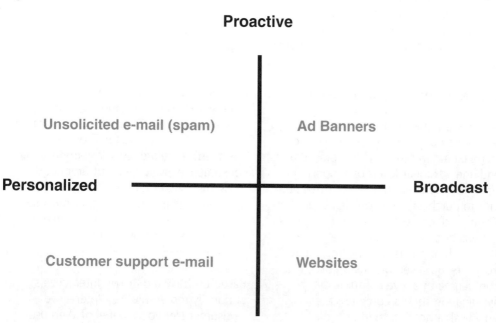

How Internet Advertising Works

first define business objectives and then assess how the technology can be used to achieve the objectives. Do not let the technology define your strategy. Some objectives are best achieved with a broadcast message, while others are best achieved with a personal touch. Some objectives require waiting for consumers to initiate the dialogue so that the brand can react to their perceived needs, while other objectives are best achieved when the brand proactively communicates with the consumer. We suggest that development of realistic objectives begin with an evaluation of:

- the nature of your product or service
- the current level of Web site category use by your target population
- perceived legitimacy of personalized communication related to your product in the minds of consumers

Exploring these areas will help the marketer develop the appropriate objectives and optimize the online marketing mix of communication options.

Nature of the product or service

Certainly the nature of the product should have a dramatic impact on the types of online communication you employ. Brands requiring 1) higher levels of support and customer service, 2) higher consideration prior to purchase, or 3) online acquisition of the product are among the categories that will benefit from a comprehensive Web site that reacts to consumers and prospects. But even brands that enjoy a high degree of active consumer investigation should communicate proactively - not just to inform the consumer that a corporate Web site can be accessed with a "Click Here" but also to carry the brand message directly to the consumer, communicating the unique brand proposition. Brands that can only be purchased offline, that have lower levels of formal information gathering prior to purchase, or that do not require significant customer support do not require a mega Web site to satisfy a consumer's need. While brands finding themselves in this category may attempt to create content appealing enough to entice users to their Web site, they run the risk of confusing the consumer regarding the focus of the brand. These brands will often achieve a greater impact on their bottom line with advertising and exclusive content sponsorships on ad-supported content Web sites than with a colossal corporate Web site.

Consumer level of Web site category use

How many of your consumers are seeking out your brand on their own accord? As Jeremy Bullmore, from the Board of Directors for WPP Group, plc, and formerly the Chairman of J. Walter Thompson observed, there are some categories where consumers actively seek out product messages (such as automobiles) and other categories where the advertiser must seek out the consumer (such as packaged goods). This is as true online as it is in traditional media.

Based on Millward Brown Interactive's nationally representative "Interactive Consumer Network" panel of Web users, we find that higher information complexity categories and categories that can be purchased online have, predictably, higher levels of Web site category use. Lower information complexity categories and those that cannot be purchased online, such as packaged goods, have lower levels of Web site use. What if your brand category has lower Web site use, but a profitable segment of your market is online? Lower Web site usage categories, such as packaged goods, can reap significant value from online marketing communication without a significant Web site investment. They can create value by focusing their online communication mix almost exclusively in Web advertising, exclusive long-term sponsorships, and similar brand communication strategies that leverage the power of the medium to proactively reach consumers and enhance the brand through advertising messages.

A tightly focused micro Web site combined with extensive brand advertising may be the most effective communication mix in these cases. Can brands reinvent the consumer relationship with the category? Perhaps. The Web can profoundly change product information gathering, purchase, distribution, fulfillment, and customer support. However, building a Web site with attractive content and a novel approach to obtaining the product is not enough. The service has to provide real value, let consumers know the service exists, and provide a compelling reason to use it. Many people have assumed that because each Web site is only a click away that the Web is a level playing field and consumers will seek out your brand. But The Web is not a level playing field. Usage is separated by awareness (at least) and often both a perceived need for your brand and preference for your service over competitive offerings. A direct marketing approach may not be sufficient to achieve marketing objectives. E*Trade's "Someday we'll all invest this way" advertising campaign provides a good example of Web advertising that communicates a brand message. While some users may click on the banner so that they can "invest this way today" through their online financial services, the ad's primary effect is likely to be enhancement of brand presence and perceived relevance in the minds of the consumers - thereby creating long-term brand health. Finding the right mix between a Web site that reacts to a consumer's needs and proactive banner advertising is key to a brand's online marketing success.

Legitimacy of personalized communication with your target

The nature of the product and the consumer level of Web site category usage help marketers determine the proper mix among proactive and reactive online marketing elements. But what about the dimension of personalized versus broadcast communication? When considering this dimension, it is important to bear in mind that consumers have finite time and can manage a finite number of relationships. Ask yourself, "Why would a consumer want to have a relationship with my brand?" Some marketers have sidestepped this question and answered, "We will require the consumer to give us their contact information - then we will have a

How Internet Advertising Works

relationship." But a relationship is more than an email or postal address in a database.

Some brands lend themselves to relationships with consumers. In these cases, personal dialogue between a brand representative and the consumer is both welcome and provides advantage in some categories. For other brands, there is no perceived need for a relationship in the mind of the consumer and personal dialogue lacks credibility with the target market - in fact, it may even annoy them. This principle is captured in the direct marketing riddle that asks: "What is the difference between junk mail and personal mail?" Answer: personal interest. Here are some basic questions to ask:
- What is the personal interest (not gimmick) for the consumer?
- What will your brand do for the consumer to justify the relationship?
- Can you maintain this relationship profitably?

Brands can communicate with different degrees of personalization, ranging from pure personal communication, such as Auto-by-tel's personal email response from a broker regarding the price and availability of an automobile that suits a consumer's request; to segmented communication like CNN's custom news service; to undifferentiated broadcast communication, which is the approach used by the majority of Web sites. Unless the marketer will be creating genuine value for the consumer - and can do so profitably - the returns from sophisticated database and dynamic delivery tools probably do not justify the investment. The return on investment can and should be tested against your economic objectives. Auto-by-tel can measure the profit from selling cars with personalized response. CNN can measure the value from incremental exposure to advertising they sell as a result of the custom news offering.

There are several powerful tools that enable mass customization of communication. The central questions related to the deployment of these tools are: What value can be created for the customer? And what is the return on investment for the brand?

Now What?

Now that you have identified the economic value of Web users and formed an appropriate online communication mix strategy for your brand, you need to make sure you have a strategy to measure your performance and gain the learning necessary to evolve your marketing strategy. It shouldn't come as a surprise to the reader that we must be certain to measure the right metrics to provide an answer to performance accountability. Measuring the wrong metrics will lead you astray. But even more important than accounting for your investment of marketing capital is institutionalizing learning that will advance and evolve your brand's intellectual capital regarding effective use of this burgeoning medium. This is what I call the Two A's of research: Accountability and Actionability.

What is the key to the Two A's? Measuring, to the best of our ability, consumer response to your marketing activities and diagnosing the improvements

How Internet Advertising Works

that should be made. In traditional media this objective is accomplished by constantly tracking brand performance with continuously fielded surveys that measure and monitor consumers' attitudinal and behavioral responses to the brand. Marketers use this powerful brand tracking tool to diagnose and enhance brand performance. For example, marketers can readily measure how a TV ad impacts the following:

- Relationship between advertising exposure and the effect on dimensions such as
  - Ad awareness
  - Brand imagery
  - Purchase intent
- Cost efficiency of the communication
- Impact of flighting and media weight on the brand

By talking directly with consumers through a structured survey, marketers have the ability to link the timing of key marketing events with changes in consumer attitudes, brand perceptions, and purchase behavior toward the marketer's brand.

Millward Brown International pioneered this research approach more than two decades ago for traditional media. The approach is used by leading brands such as Kraft, GM, Hewlett Packard, Disney, and Levi's. These brands use this information to help them make crucial decisions related to brand management and communication.

Recently, Web advertisers have gained the ability to gather the same feedback and learning from online brand communication. These new research capabilities can measure all major online marketing activity - ranging from Ad banners to a targeted Web site communication. The power of the insight is substantial. Brands that had previously relied on impressions, clickthroughs, and faith that their Web advertising works (proven in general terms) now have the type of specific insight that allows them to tangibly measure the value of everything from specific elements of their Web site to the differential performance of specific Ad banner executions. This new research system works by matching the marketing activities with consumer response (gathered through a survey with a brand's online target audience). In fact, brands that currently have the brand tracking research program in place for traditional media can compare the value of key elements of their online marketing to traditional marketing communication (such as television and print advertising). This "across media" insight allows brands to understand how they can improve the return on investment from their communication mix both online and offline. This new type of Web research insight can provide answers to the following difficult questions:

1. How does my target market perceive my brand versus competitive brands?
2. What creative is likely to work best for specific target markets?
3. What Web sites are most efficient at reaching my target audience?
4. What is the audience profile behind the ad impressions and clickthroughs?
5. How have advertising and marketing activities impacted my target audience?

To date, brands operating online have lacked the insight provided by rigorous research that focuses on addressing these questions. Yet, these questions are central to a brand's success. Marketers now have research tools within their reach that provide accurate measurement of the impact of their communication - from mass-reach television to targeted Web advertising to a narrow-cast Web site. These tools empower marketers to determine how each of their brand communications affect consumers as well has how each component interacts to ensure they are achieving their objectives. This feedback loop helps marketers make even more effective use of the online marketing vehicle.

As a result of solid online strategy and new online market feedback tools, brand managers can confidently address the complex questions of what is (and is not) working for their brand. Applying these tools makes marketers more effective and efficient - ensuring that they have an accurate accounting of results and actionable learning to build their brands into the future.

It Pays To Advertise. Effects of Business-to-Business Advertising on Decision-Makers: Results of Recent Research

Title: It Pays To Advertise. Effects of Business-to-Business Advertising on Decision-Makers: Results of Recent Research
Author: American Business Press
Abstract: The cost per contact of an advertising view in a specialized business publication is 32 cents. Getting an Internet contact costs 98 cents. A bulk mail direct mail costs $1.68 per contact. The other methods are far more costly: from $13.60 for a business letter to $277 for a sales call. In other words, the cost per impression of a B2B ad is less than 1/800 of the cost of a sales call.

Copyright: © 1999 American Business Press. All Rights Reserved.

Introduction

An important part of the marketer's task is to select the right medium, form, and frequency for advertising contacts with existing and potential buyers. The center of the marketing target is the business decision-maker who has a current product need and can respond to an effective message by placing an order. The next best case for the marketer is a campaign that builds the brand in the minds of buyers and specifiers who don't have a current need, but will remember the message when a purchasing need arises. Developing customer loyalty is extremely important: it costs 5-10 times as much to attract a new customer as it does to make repeat sales within an installed customer base.

Today's market is a battle for "Brandwidth": which might be defined as getting attention for a particular product and its message. Current research shows that print business publications are an efficient and cost-effective way to influence buying behavior. On the average, managers and professionals read 7.6 specialized trade magazines, for a total of 18 working days a year. In fact, 44% of this group expect to increase the time spent reading professional magazines over the next few years. Only 11% think they'll devote less time to this task. Thirty percent consult a specialized business publication at least once a day; 44% do so two or three times a week.

The cost per contact of an advertising view in a specialized business publication is 32 cents. Getting an Internet contact costs 98 cents. A bulk mail direct mail piece costs $1.68 per contact. The other methods are far, far more costly: $13.60 for a business letter, $31.16 for a telemarketing contact, and $277.00 for a sales call. In other words, the cost per impression of a B2B ad is less than 1/800 of the cost of a sales call.

Establishing and Defending Brand Equity

According to marketing experts, a company's brand is the expression of an

It Pays To Advertise. Effects of Business-to-Business Advertising on Decision-Makers: Results of Recent Research

idea through all marketing elements. Every marketing activity, in traditional media or online, either adds to or subtracts from brand equity.

Building the brand works in good times and bad. A company selling to capacity still must build and maintain a positive image that comes into play when demand weakens. The function of the ad campaign is to communicate the positive values of the brand to readers. If brand recognition dips, the message must be imparted to precisely targeted potential buyers.

There are four steps in building a brand:

1. Create AWARENESS of the product or service; familiarize existing and potential customers with brand attributes such as innovation, value, customer service, trust, and technology leadership
2. Induce CONSIDERATION of the products and services by those exposed to the advertiser's message
3. Establish PREFERENCE for the brand by stressing its values to the customer, and so they will spread the message by recommending the brand
4. Inspire PURCHASE (preferably with up-sells, cross-sells, and repeat business).

The recent studies demonstrate that the best strategy is a coordinated marketing campaign that makes full use of all appropriate media, including business to business print media, trade shows, directories, the Web, and perhaps newsletters and data bases.

Specialized Print: The Trusted Medium

Our contention is that business to business print media have proved their centrality to the marketing campaign. Purchasing decision-makers find B2B publications to be the most informative, most credible source of knowledge about products. In these publications, for this audience, ads are not an annoying distraction: they are a welcome guide to needed purchases.

In a December 1998 survey, respondents gave trade journals a "Media Credibility Index" of 350. This was by far the highest rating of any communications medium. The Economist scored 320; the New York Times was rated at 254, and Wall Street Journal at 230; Business Week scored only 160. CNN rated only 129, CNBC only 112.

Business decision-makers think that industry trade publications are objective: 75.2% of them call the trade press "highly objective" or "somewhat objective." National daily newspapers got an objectivity rating of only 60%, and general news magazines were rated somewhat or highly objective by only 55.5% of respondents.

When asked which media were "extremely useful" or "useful" to them, the vast majority of respondents found business magazine editorial content and business magazine ads helpful. In fact, the usefulness scores for business magazine ads went up significantly between 1996 and 1998.

It Pays To Advertise. Effects of Business-to-Business Advertising on Decision-Makers: Results of Recent Research

The articles and ads in specialized business publications rated higher than catalogs, exhibits at trade shows, visits from reps, direct mail from the vendor, or Web sites (although Web sites gained a lot of credibility in two years). Ads in specialized business publications were considered useful or very useful over three times as often as advertisements in generalized business publications that are not tailored to a particular industry.

On a scale of 1-5, where 1 was ranked "very low" and 5 was "very high," specialized business publications got a rating of 4.0 from business decision makers asked to choose the "most informative" medium. This was by far the best result: directories scored 3.48, direct mail scored 2.66, radio scored 2.43, and television scored 2.28. Furthermore, specialized business publications rated as the least annoying medium (1.86); fax ads scored 3.74 as "most annoying," followed by TV at 3.42 and direct mail at 3.37. So specialized business press advertising provides a lot of positive feeling, and very little negative feeling.

Not only do people in business think of the business press as trustworthy, useful, and informative, they find it the most helpful medium for doing their jobs. To 41% of managers and executives, business or trade periodicals were the most important medium for doing their jobs. About a quarter (29%) placed the highest value on personal contacts, but only 10% gave first place to general-readership daily newspapers, 9% to broadcast news, 7% to the Internet.

Utilization of Business Media

Research shows that business owners, professionals, and executives with purchasing responsibility spend a significant amount of time interacting with trade journals; that they have several objectives in using the trade press; and that they find contacts with business to business print media interesting, useful, and helpful.

More than one-third of buyers and specifiers for B2B purchases look at professional magazines first when they want to learn their options for procurement for a new project. In 1998, the nearest competition for this role came from print directories and reference books, and broad-based search engines, which scored only 15% each, making specialized business magazines over twice as effective. In 1998, only 12% thought first of talking to the sales rep—a serious decline from 1997, when 20% made this their first step in the procurement process.

In addition to devoting much time to reading business journals, buyers and specifiers clip and file stories and advertisements that they find particularly appealing. It is very unlikely that they would record and refer to broadcast advertisements, so the retention factor is a major advantage of business print media. Nearly all (98%) of trade journal readers consult their clip files at least occasionally. One-third (34%) do so at least once a month, and the average number of annual references is 10.8, or approximately once a month.

It Pays To Advertise. Effects of Business-to-Business Advertising on Decision-Makers: Results of Recent Research

More than half of business press readers state that they often find information in those sources that is not available in other media, and 45% sometimes find that the business press has exclusive information. The following table shows business press features enjoyed and found useful by readers.

intensively as they did five years ago, and in fact 35% view ads more frequently than five years ago. Eighty-four percent look at the ads in the professional journals they read at least 50% of the time, and 52% always or almost always inspect the advertisements.

Table 1

Valued Features of Business Press	%
Portable	73
Convenient	67
Articles and ads can be clipped and saved	58
Easy to save for reference	31
Can be highlighted and written on	31
Can be adapted (folded, dog-eared)	20

Advertising in specialized business publications affects other media as well. It influences which trade show booths will be visited, and induces potential purchasers to contact sales representatives. The representative's job is much easier if the contact is already familiar with the company and its products, and views them in a positive light. Advertising is also a major driver of traffic to corporate Web sites.

Although impressions are greater in general-interest business publications, the disadvantage is that most of the publication's readership will fall outside the advertiser's particular industry.

Positive Response to Advertising

Virtually all business buyers and specifiers (94%) look at business magazine ads as

Readers of the specialized business press don't find ads an annoying interruption in their reading experience: quite the contrary. Close to two-thirds (65%) find advertising content a very important or important part of the overall presentation of a business magazine, and 30% find it somewhat important.

About half of business buyers look at trade magazine ads more frequently than they did five years ago. This table expresses what readers like about the advertising in business-to-business publications (as distinct from what they value about the publication as a whole). 77% use trade-publication ads to learn about new technology.

It Pays To Advertise. Effects of Business-to-Business Advertising on Decision-Makers: Results of Recent Research

Table 2

Appreciated Features of Business to Business Advertising	%
Alerts reader to new technology	96
Ability to learn about new vendors	76
Ability to send for additional information	83
More accurate impression of individual vendors	47

Business managers and administrators (M&A), and purchasing managers and agents (PM) take action in response to business advertising.. This chart shows ad-related behaviors for the twelve months before the survey. High percentages see an interesting ad, then contact the advertiser for more information (CA); file the ad for future reference (FA); talked about the ad with a colleague (TA); routed the ad to a colleague (RA); or requested a price quote (PQ).

Over half of purchasing managers and agents (50.2%), and 41.1% of managers and administrators, asked for a sales call based on an ad, and a small majority (55.7% of managers and administrators, and 52.9% of purchasing managers and agents, went ahead to complete the purchase from the advertisement.

The more of a specialized business publication's editorial content a business decision-maker reads, the more ads he or she is likely to read. People who read more than three-quarters of a publication's editorial content viewed 1.6 times as many ads as the average.

The Virtuous Circle: Branding on the Web

Obviously, the World Wide Web contains billions of pages of information, on millions of sites, and the amount of information grows every day. Major search engines only find about one-third of all sites, and it's easy for a particular corporate site to get lost in the crush. It's important for the URL of the company's site to be integrated into its marketing campaign, especially its print ads.

Traditional and new media can interact in a mutually productive manner that increases corporate sales significantly—a phenomenon nicknamed "The Virtuous Circle."

Interest is developed through print, trade shows, seminars, Point of Purchase, broadcast, and other rich media. As noted above, it is very helpful to feature the URL of the corporate site throughout the marketing campaign. In this paradigm, intention to buy and customer involvement are developed online. The business' Web site is responsible for service branding and service marketing. Customer loyalty is developed by providing solutions for customers. Online users go to magazine Web sites an average of 2.4 times a week.

It Pays To Advertise. Effects of Business-to-Business Advertising on Decision-Makers: Results of Recent Research

The strategy is complemented by effective use of e-mail, the interaction medium between business and customer. The site must make it easy for customers and potential customers to make contact, initiating a one-to-one dialogue. E-mail also works powerfully as a retention medium permitting the seller to send reminders (when it's time for a re-order) and inform customers about other products they might find useful.

This table shows results of research on Cahners publication readers.

Table 3

Value	Agree Strongly	Agree Somewhat	Total
Access to information they would not search for on Internet	55%	37%	92%
Learn about manufacturers' Websites	37%	43%	80%
Reading is more enjoyable than Web surfing	37%	36%	73%

More than half of surveyed Cahners readers say they go to Web sites based on the content of specialized B2B publications (56% from ads, 53% from articles). These are by far the most potent drivers of Web traffic in this market: 41% got URLs from colleagues and word of mouth, 40% got links from the Web sites of business publications, and 28% got them via direct mail.

Early in the purchasing process, trade magazines create awareness, give readers positive perceptions of your company, and create leads. At the middle stage, trade magazine sites are highly valued; at the late stage, buyers/specifiers typically turn to vendor Web sites for more detailed specs.

When researching potential future purchases, 91% of buyer/specifiers consult trade magazines (although this percentage drops to 56% in the middle stage, and 38% at the advanced stage when a purchase is impending).

Establishment of contact

In 1998, 84% of B2B marketers surveyed advertised in specialized business publications. This was one of the most popular methods used, tied with trade shows and the Internet and other electronic media. 74% used direct mail, 70% used publicity and public relations, and only 47% advertised in general magazines.

As we recommend, respondents made B2B advertising central to their market plan. It attracted the largest share of the budget (23% in 1998), versus 18% for trade shows, 10% for direct mail, and only 9% for the Internet. More than half (55%) planned to maintain their spending level on B2B advertising in 1999; 18% planned to cut back, and 27% to increase it.

In the sales partnership, contact usually comes from the buyer first: 82% buyer-initiated contacts, versus 18% initiated by a sales representative. Almost all buyers get

It Pays To Advertise. Effects of Business-to-Business Advertising on Decision-Makers: Results of Recent Research

their first information about a vendor's products through the vendor's ads and other communications. Only 6% of initial contents come from sales reps. But that's just fine as far as customers are concerned. Nearly all (95%) want to have at least some information about the vendor before they see the rep, and over nine-tenths (91%) are more likely to let the sales representative get a foot in the door if they have at least some general knowledge about the vendor company.

For 39% of respondents, trade magazines were the first medium consulted about finding new technology alternatives; 20% went first to sales representatives, and only 1% went to general business publications.

Generation of Brand Awareness

Additional Cahners research shows that the primary interest of business press readers is news about new products, equipment, systems, and other aspects of technology. More than two-thirds of them (79%) named this as one of the two most important benefits of reading professional magazines. Over half (52%) want brief articles on the application of specific technologies; 33% want how-tos, 16% want industry news and 7% want news about suppliers.

Table 4 shows the reaction of customers to a well-done continuing advertising campaign..

When readers see larger, more colorful ads from a particular vendor more often, it influences their impression of the vendor. Two-thirds (68%) think that the vendor must have ample funding, staffing, and other resources. One-third (34%) think the vendor must be a market leader. One-fifth (22% and 19%) think it must be more experienced or provide better support or service than its competitors.

Development of Interest

Where do companies get their best leads? 44% of respondents told Cahners that their best leads come from print advertising; 31% said the Web generated the best leads, and 25% found both equally valuable. In other words, about two-thirds of marketers found that print advertising produced high-quality leads.

Table 4

Perception of Company	%
It is well-established	57
Likely to have broader product line than competitors	45
One of the most successful companies in its market segment	42
Likely to innovate	38
Likely to have broad product line	25
Likely to have above-average quality	24
Likely to have above average reliability	30

It Pays To Advertise. Effects of Business-to-Business Advertising on Decision-Makers: Results of Recent Research

Preference Building

In the past two years, customers have changed somewhat in the way they respond to advertisements. Calling an (800) number is still by far the most popular response, used by 71% of respondents (versus 80% in 1997). Currently, 68% mail or fax a reader service card, versus 73% two years ago. Close to half (46%) contact a sales representative as opposed to a 58% contact rate in 1997.

The big gainers have been e-mail responses to the advertiser (28% in 1999, 17% in 1997) and going to the advertiser's Web site (56% versus 25% in 1997; more than double the rate).

Offering Specific Proposals

Survey respondents were asked to choose the three major benefits they valued most..

Table 5

Information Most Valued by Potential Customers	%
Availability/delivery date for merchandise	51.0
Price list	45.2
Product specifications	37.7
Complete catalog	36.9
Customer service/tech support information	32.7
Detailed quality/reliability information	21.4

Closing Sales

Readers are usually satisfied with the amount of information they get when they ask for vendor information (93% got the information they asked for), and the information arrives in time to answer the reader's questions 98% of the time. However, only 30% of those who send in reader service cards are actually contacted by a sales representative in person or over the phone, demonstrating that there is significant potential to close additional sales, and improve customer loyalty, by stronger follow-up on leads.

Keeping the Customers Sold

Even after they buy, business specifiers almost universally stay in contact with their vendors' ads. 94% read their suppliers' ads, to see the new models and features, how the vendor's image and operations are changing, to keep up with upgrades, bug fixes, and support programs; and to get reassurance that the vendor was a good choice.

Customers are comfortable with stable relationships. Over three-quarters (77%) have lists of preferred suppliers, and 57% have formal partnering relationships. About two-thirds of respondents say they like these long-term supplier "marriages" because they promote sharing of information, on-time delivery, and consistency of product quality.

CONCLUSION

Now and in the foreseeable future, the business press is the "first-read" medium for business purchasers and decision-makers. Research shows that specialized business publications lead in convenience, credibility, and perceived information content. To satisfy goals such as attracting attention, brand building, and creating and rewarding loyalty, the marketing campaign should start with and center on specialized business publications.

The Seven Steps to Successful Direct Marketing

Title: The Seven Steps to Successful Direct Marketing
Author: Carey Hedges, HN Marketing Ltd

Copyright: © HN Marketing Ltd 1999
Biography: Carey Hedges is the owner and managing consultant of HN Marketing Ltd. Technically competent, with many years practical marketing experience, HN work only in the IT industry. Their capabilities cover many areas of communications from copy writing to full campaign management. They have used their technical understanding and knowledge of the market to help many organisations generate effective sales leads, improve customer communication and raise their profile in the industry.

Introduction

Are you in any doubt that direct marketing is growing in popularity? Judging by your own in-tray each morning - probably not. The average busy executive receives in excess of 60 items of mail per day (including the electronic variety). In the 1970's direct marketing in the business to business field meant little more than catalogues and mail order. Today British businesses send a staggering 760 million pieces of direct mail per year, accounting for 32 percent of all British direct mail. (Sorry rest of the world, no figures were available at the time of going to press.) So why is this? What has fuelled this new love affair with the mail, and should you be joining the bandwagon?

The problem with direct mail is that, on the face of it, it appears deceptively simple. Send a letter and the world will flock to your door! With average response rates running at around 2%, we marketeers might well be forced to revise the old adage…I know that 50% of my marketing budget is wasted - I just don't know which 50%. Only 50%? Count yourself lucky!

This snipe is perhaps a little flippant because one of the three main benefits of direct mail is its measurability. It is easy to test different formats and different lists. You can quickly know if a campaign is working or not and adjust your investment accordingly. Done properly, direct marketing shines a very bright light into the corners of your marketing budget and enables you to measure your return on investment. Targetability and flexibility are the other two benefits.
Targetability because you can focus your message to as large or small a group as is appropriate, and flexible because you can batch the distribution for easy follow-up or mail en masse for a fast response.

But the key to success is to do it properly. What make the difference between good direct mail and bad direct mail? No doubt as a recipient of direct mail you will have your own opinions.
For my money there is one overriding

The Seven Steps to Successful Direct Marketing

factor that makes the difference between rubbish and response…..Relevance. Write it down. Remember this word. Your mailer won't work without it. Although the rest of this article is devoted to the logistics - the seven steps to implementing direct mail, there is only one key to success: your message must be relevant to the recipient. If it's not, then your mailer will go to the large round receptacle at the side of his desk.

I have seen a lot of bad direct mail - including the all time winner…. One with no response mechanism at all. I suppose that the sender intended to follow up by phone. One way or the other I never got the call. Is this what they call a 'teaser' campaign I wonder? I have also seen a lot of good direct mail. Timely, relevant, an enticing offer, a simple mechanism for reply, - It's not so difficult really.

Here are my seven steps to successful direct mail

1. Clear Objectives

The most important step is to have clear objectives. It may sound a bit too 'textbook', but it is
true. If you understand what it is you are trying to achieve then the rest is usually a logical progression.

Theodore Levitt, the famous marketing guru, describes the process of direct marketing as 'making and keeping customers'. Of course he is referring to the combined forces of telephone and post but the sentiment is equally applicable to each activity alone.
It is all about building relationships - and it is therefore sometimes called relationship marketing. It speaks (writes) directly to the individual. It should demonstrate an understanding of the individual and establish a dialogue between the seller and the individual. Based on this level of individualism, would it not therefore be impolite and be seen as time-wasting to promote a range of Solaris support services to a DEC site? Relevance again. Direct mail can take on a variety of objectives from newsletters aimed at maintaining a relationship, to the 'cold call' aimed at striking new ones. The tone and content of the message will be quite different in each case and so your goals must be clear if the communication is to meet them.

2. Know Your Audience

Know the people with which you need to communicate. What are they interested in? Where are they based? What problems are they trying to solve? In order to make your message relevant you must understand your target audience.

This might mean that you need to do some research to build a profile. As a minimum you should know the following about the people you want to mail:
- Industry sector or the application that is driving the need.
- Geographic location.
- Size of organisation in terms of number of employees or turnover.
- The decision-maker or the composition of the decision making team.

At the very least this information allows you to eliminate the no-hopers. For

The Seven Steps to Successful Direct Marketing

example, the companies that are too small to be able to afford your solution or those that have no need.

There is not much point, for example, offering 'state of the art software to streamline your production line' if the company has no production line.

Having profiled your audience, there are several ways to get a list of contact addresses: directories, list brokers, magazine lists etc. Ask your colleagues in the industry to recommend a supplier that they have used. Word of mouth is often the best way to identify a quality list. If you need some help to get started, HN Marketing would be happy to help. Alternatively, the European Direct Marketing Association can put you in touch with list owners and brokers in the different European countries. Their address is at the end of this document.

3. A Relevant Message and Offer

OOPS that word again. What makes your solution, hardware, software or service attractive to your target audience? What sets it apart from the competition? This is its USP or Unique Selling Proposition. USP doesn't mean that your product necessarily is unique, but that a combination of product, availability, price, packaging and service makes it uniquely desirable to your audience – a SMSP, single-minded selling proposition. Beware enthusing about the 96 Terabyte capacity of the TLA architecture… Explain why this means that your target might need one. Think benefits, not features. If your prime objective is quantity, perhaps because you

are building a list for farming in the future, then offering a gift or a prize will certainly boost your response rate. You will receive more enquiries, but a good portion of these individuals will have responded because they wanted the gift, not because of their interest in your solution.

Gifts and prizes are classed as sales promotions and there is a strict code of practice that you need to be aware of. Copies of the code of practice and other advice is available from CAP (Committee of Advertising Practice) in the UK. The ASA (Advertising Standards Authority) is a founder of the European Alliance that monitors adherence to good practice on a European level. Addresses for both of these organisations are at the end of this document.

4. Simple Design With a Clear Call to Action

What is the response you would like your audience to take as a result of receiving your message? - Be realistic now… Do they have enough information to make a purchase decision or is it too early in the relationship?

Usually you want them to call, fax or reply in some way to show their willingness to enter into a dialogue with you. The easier you make it for them to respond the more replies you will get.

Even in this modern age of communication, fax-back or postal replies are still more popular than phone.

In the UK, the Royal Mail offers a number of pre-paid response mechanisms. Other post offices around Europe have equivalent services. The difference in rate-of-return between sender pays vs. recipient pays response is most marked in the consumer sector. The difference is quite small in business to business mail. Pre-paid replies do however remove the need for a stamp or franking, and so bring the response one step closer.

FREEPOST allows you to pre-pay replies that come from within your country. IBRS (International Business Reply Service) lets you pre-pay for European-wide responses. There are specific design regulations to adhere to in both cases, including the weight and thickness of the response card. You can find out more by contacting the Royal Mail in the UK or your own national postal company, or talk to your marketing agency.

The jury is still out on whether appearing 'local' is the best bet or whether the exotic appeal of the overseas supplier can be more enticing. If looking local is important to overcoming any cultural resistance then services such as Admail can help. The Royal Mail (and others I'm sure) can supply you with an in country PO Box and advise you on the local format for the address. A discrete code within the address allows your UK address to be identified and responses to be forwarded to you.

A word on localising your message. A mailer using the first language of your target audience will have a better response than one that doesn't. Depending upon the size of each geographic territory you may decide it is worth making the investment in translation. Make sure that the translation is prepared by someone who is fluent in that language. Usually this means that they are a native speaker and still live in the country. Languages evolve continuously and are subject to trends and fashions. A six-month absence from the country is noticeable to a native. Also, if your message is technical make sure that your translator knows the local terminology.

Beware too the use of puns and clever plays on words. They often demand a grasp of a second language that is beyond your average reader, or they just don't translate.

Design can also play a part in reducing cost in a multi-lingual piece. Keeping all the text in black means that you will only have to change one printing plate between versions. Lighter weight paper can save on postal charges. Ensuring that any piece can be machine-inserted into an envelope will reduce fulfilment and handling costs. Let the design agency know exactly how the piece is to be used and any restrictions that they need to take account of.

6. Fulfilment - Consider Your Options

In any campaign, postage is the single biggest item of expenditure. Frightening isn't it? You have spent all this money to produce the mailer and now you have to spend the same again to mail it out. Money can be saved. Consider your options.

The Seven Steps to Successful Direct Marketing

Surface mail is the most economical 'standard' service and will deliver your piece to its destination within 2 weeks. Airmail in comparison, will deliver to European cities in 3-4 days.
If you need a faster service then you are into paying for courier or express delivery. If you are mailing in volume then bulk discounts may be available. Talk to your postal service provider to find out what is on offer.

In the UK and for mail destined for a UK address, Mailsort discounts are offered on Large mailings in excess of 4000 letters. The discount can be 13% or more, depending upon the service you require - Mailsort 1 is for next day delivery, Mailsort 2 for within three working days and Mailsort 3 for within seven working days. Each address should have a postcode and the letters need to be sorted in accordance with the Royal Mail's Mailsort database. For overseas and European mailings, Airstream offers attractive pricing based upon the weight of the consignment (not each piece) and the number of pieces. Air stream offers the speed of airmail. For mail to EC countries a customs declaration is no longer required for letters and small packets. You should check out the customs requirements for mail to other destinations.

half of the story. You have to maintain the dialogue to make the sale.

Modern databases make it easy to capture information about your suspects. Next time you want to mail them en masse, there they will be, only a few key strokes away and you can have mail-merged letters en route with the information they asked for.

Modern databases also let you track your responses. A discrete code on the reply coupon will let you know which list the responder is from, when the mailer was sent and, if you are testing a headline or offer, which version of the mailer the response is to. By recording any leads that convert to orders you can measure the effectiveness of each variation of the campaign. Such numbers can ensure that you learn what works and what doesn't for your target audience, so building a stronger relationship with your customers and prospects.

7. Follow Through

Don't leave your hard-earned responses languishing in a dusty corner of the office waiting for someone to sort through them. The volume of responses may be excellent and the boss may be impressed with your campaign, but getting the response is only

Useful Sources of Information

You may find information from the following sources helpful in your campaign planning:

European Direct Marketing Association
36 Rue du Gouvernement Provisoire
1000 Brussels
Belgium
Tel: +32 2 217 6309
Fax: +32 2 217 6985

Advertising Standards Authority (ASA)
22 Torrington Place
London WC1E 7HW
Tel: +44 (0) 171 580 5555
Fax: +44 (0) 171 631 3051

Royal Mail
0345 740740 Customer Services
0345 950950 Sales centre
Royal Mail International HQ
49 Featherstone Street
London EC1Y 8SY.

HN Marketing Ltd can be contacted as follows:
Tel: +44 (0) 1628 622187
Fax: +44 (0) 1628 672813
Email: info@hn-marketing.co.uk
Web: http://www.hn-marketing.co.uk

Exploding the WEB CPM Myth

Title: Exploding the WEB CPM Myth
Author: Rick Boyce
Abstract: There's a lot of chatter about Web cost-per-thousand (CPM) calculations. Can you compare Web ads with television ads? The truth is that for reaching high-income, well educated audiences, television CPMs are quite high, and Web advertising, in many cases, can actually deliver a lower CPM against the same demographics.

Copyright: ©1999 Internet Advertising Bureau. All Rights Reserved.

If you've read anything in the media or from Internet analyst reports recently, you've no doubt noticed all the chatter about Web cost-per-thousand (CPM) calculations. The common theme seems to be that Web CPMs are too high, particularly compared to television. Take for example this comment from Morgan Stanley's industry bible, The Internet Advertising Report: "While CPMs on the Web vary widely, on average, they have been at higher levels than they are in most other media."

Much of what is published by the analyst community promulgates the Web CPM myth. As any media planner will tell you, it is impossible to compare CPM objectively across different media without accounting for the level of targeting the advertiser wants to reach. The chart on the next page, from Jupiter Communications, appeared prominently in the Morgan Stanley Internet Advertising Report. Often quoted but typical of many CPM analyses that are generated, some may feel this chart can only possibly intimate one conclusion: that Web CPMs are 66% higher than TV CPMs. But perhaps, since these two types of media are so diverse, a slightly different type of comparison would, in fact, be possible.

What is important to remember when looking at comparative CPMs is that TV is a mass reach medium. Even selective dayparts and programs reach giant cross-sections of the population. While the thirty-second network primetime spot referenced in Exhibit I delivers a $12 CPM against all U.S. households, if the target audience is refined by adding income and education screens, the CPM jumps dramatically. The truth is that for reaching high-income, well-educated audiences, television CPMs are quite high, and Web advertising, in many cases, can actually deliver a lower CPM against the same demographics.

Table 1

Exhibit I: Ad Rate Comparison Across Major Media

MEDIUM	VEHICLE	COST	REACH	CPM
TV	:30 network primetime	$120,000	10 million households	$12
Consumer Magazine	Page, 4-color in Cosmopolitan	$56,155	2.5 million paid readers	$35
Online Service	Banner on Compuserve major topic page	$10,000 per month	750,000 visitors	$13
Website	Banner on Infoseek	&10,000 per month	500,000 page views per month	$20

Source: Jupiter Communications

Using network primetime costs and ratings estimates from the fourth quarter of 1997, Exhibit II examines the reach and CPM of an average primetime when age, income and education screens are added.

Table 2

Exhibit II: Network Primetime Reach and CPM Vs. Select Demographics

TARGET DEFINITION	AUDIENCE ESTIMATE	CPM (CROSS)
US TV Households	8,526,000	$12.52
A18-49	6.361,000	$16.78
A18-49 & $40k + HHI	3,796,000	$28.12
A18-49 & $40k + HHI & HOH 1 + yr. College	2,171,700	$49.15

Key: HHI = Household Income, HOH = Head of Household
Source: Netcosts Q4 1997 Primetime Scatter Estimates; Nielsen 11/97 Audience Estimates

Exploding the WEB CPM Myth

Now, let's do the television/Web CPM comparison, and let's assume that our target audience is college educated and enjoys an above-average income. If we adjust the primetime CPM from $12 to $49 to account for the premium required to reach an upscale target audience with TV, the television/Web CPMs actually become far more comparable. As a single point of reference, consider that a banner schedule can be purchased on Quote.com – which delivers an audience with an average household income of $110k and 70% college graduates – for under a $50 CPM. Once television CPMs are adjusted to account for reaching a specific target audience, the economies of Web advertising – particularly for reaching discreet, affluent target audiences becomes clear.

It is worth noting that online media planners and buyers can target infinite demographic or psychographic audiences because the Web is home to more vertically targeted content than any other medium. This allows advertisers to purchase media schedules against nearly any target audience with a minimum amount of waste.

But even if Web CPMs aren't out of line relative to television, is Internet advertising really effective? The answer is Yes, and here's why:

In a study conducted in 1997 for the IAB, MBinteractive (www.mbinteractive.com) concluded that "online advertising is more likely to be noticed than television advertising." The reasons that ad banners performed so favorably has to do with:

1. the lower ad-to-edit ratio on Web pages, and
2. the fact that Web users actively use the medium as opposed to passively receiving it.

With regard to the proportion of advertising to editorial, think of it this way: the typical ad banner is 468 x 60 pixels, equaling a total of 28,080 square pixels. The commonly accepted default screen size is 640 x 480 pixels, for a total of 307,020 square pixels. This means that a typical Web page is 91% editorial and 9% advertising. Dramatic when compared to magazines, which are typically in the 50/50 range, and television, which is closer to 60% programming/40% advertising.

To the point of passive vs. active viewing, television attention-level research is particularly interesting when contrasted to the online advertising effectiveness work done by MBinteractive. It's remarkable how few people are actually paying full attention to the television shows – not to mention the advertising. According to MRI, which measures television viewer attentiveness, not even two-thirds of the viewers of television's most popular shows – like Seinfeld, 60 Minutes and Mad About You – report paying "full attention" to the programming.

And those actually watching the television commercials represent even a far smaller number. A 1993 study from The Roper Organization, Exhibit III on the next page, found that just 22% of adults reported that they actually watch television commercials.

Table 3

Exhibit III: Percent of Adults Who Report They Often Respond to Television Commercials in Selected Ways

	TOTAL	FEMALE	MALE
Watch commercials	22%	19%	24%
Turn down the sound	14	13	14
Change channels	27	31	24
Leave television	45	44	47
Get annoyed	51	51	50
Talk to others, ignoring commercials	35	36	34
Talk to others about commercials	14	13	16
Get amused by funny commercials	26	24	29
Pay attention to information on new products and services	10	8	11
Fast forward through commercials when watching a taped program	35	38	32
Learn about products and services of interest	17	14	19

Source: The Roper Orginization, 1993

To contrast, you simply can't navigate the Web without high concentration and high attention levels, and both are required for advertising to get noticed, remembered and ultimately acted upon.

Exploding the WEB CPM Myth

The lowly CPM has hit the big time and is no longer destined to an obscure existence like other media acronyms: GRP, CPP, RPC, etc. But, if CPM is destined to become a mainstream term, we all need to work to understand how to use it to make accurate observations and smart analyses.

What Advertising Works?

Title: What Advertising Works?
Author: Bill Doyle, Mary A. Modahl, Ben Abbott, Forrester Research
Abstract: Advertisers on the Web have three choices. They can built: 1) destination site; 2) micro sites, small clusters of brand pages hosted by content sites; 3) banner campaigns and other low-overhead Web advertising-like sponsorships. To understand which is best, advertisers need to ask: Can it be sold online and is it a considered purchase?

Copyright: ©1999 Internet Advertising Bureau. All Rights Reserved.

Advertisers are rolling up their sleeves to figure out how to make advertising work on the Web. They are torn about what to spend and who to hire. But as Web advertising becomes a significant portion of marketing budgets, advertisers will need to adopt a set of best practices for this new medium.

Advertisers are searching for the formula that will unlock the potential of the Web. Forrester interviewed 51 companies that currently advertise on the Web and found that:

- Spending patterns vary. Consumer brands spend a small fraction of their budgets on the Web.
- Technology companies spend five times more of their brand dollars there. Spending on sites
- Still exceeds spending on ads.
- Banner campaigns run the gamut. There is no such thing as a typical Web advertising campaign.
- Ad pricing frustrates advertisers. Nearly three-quarters of the respondents want pricing to be based on results rather that CPM.
- Personalized targeting has not yet taken hold. Advertisers target mainly on content.
- Web advertising needs best practices. As Web advertising becomes a significant portion of marketing budgets over the next four years, advertisers need to develop a set of best practices around what to build and what to pay.

WHAT TO BUILD

Advertisers on the Web have three choices. They can build: 1) destination sites, which use information, entertainment, and high production values to pull users in and bring them back again; 2) micro-sites, small clusters of brand pages hosted by content sites or networks; or 3) banner campaigns and other low-overhead Web advertising-like sponsorships (see Figure 1). To understand which is best, advertisers need to ask:

What Advertising Works?

- Can it be sold online? Products that can be sold online and shipped economically or delivered digitally-such as music, tickets, books, software, and mutual funds-can use a destination site to support everything from brand awareness and consideration, through post-sales support. But if the Net doesn't enable your company to offer a product faster, cheaper, or better, rule out a destination site.
- Is it a considered purchase? Sellers of complex products like computers, cars, and industrial coatings can use a Web site to squeeze costs, allowing prospects to check specifications, configure their purchase, and get product support on-line (see Figure 2). But if your customers are more likely to ask their neighbors than you about your product, you don't need a Web site.

Destination Sites Create A New Channel

Destination sites are right only for companies that can use the Net as a full-fledged channel for exchanging information with customers, in order to book a sale (See the September/October, 1996 Leadership Strategies Report, "The Fourth Channel: Vision.") Advertisers that do build destination sites should:

- Do a gut check beforehand. A company must be willing to spend $3 million or more to build a destination site and then more to maintain it. A half-baked site actually will erode a brand. Don't underestimate the volume of customer interactions. L.L. Bean dedicates a team of customer service reps to reply to e-mail seven days a week.
- License, don't produce content. True content providers always will outpublish advertisers. When Toyota first moved onto the Web, it created seven different lifestyle 'zines. Today only Car Culture remains-and that will soon be produced by one of the big auto publishers.
- Get found. Once a site exists, marketers need to promote it. Smart companies dedicate at least 20% of their overall interactive budgets to promoting the site online. They also sneak URLs into print and TV ads and become experts in such Web guerrilla marketing techniques as getting found by search engines and trading links with other sites.

Micro-Sites Are Sufficient For Considered-Purchase Products

Micro-sites enable advertisers to communicate deeper product benefits and collect customer information without the cost of a full-blown Web site. Advertisers of considered purchases like clothes and appliances should:

- Put micro-sites where the audience is. The rumored $3 million that Levi's spent on its early way-hip site could have put micro-sites on all the top youth sites for months. Appliance manufacturers like Whirlpool should embed product selectors in such sites as remodeling.hw.net.
- Maintain clues that this is advertising. Commercial content masquerading as editorial undermines the trust on which brands are built.

What Advertising Works?

Figure 1

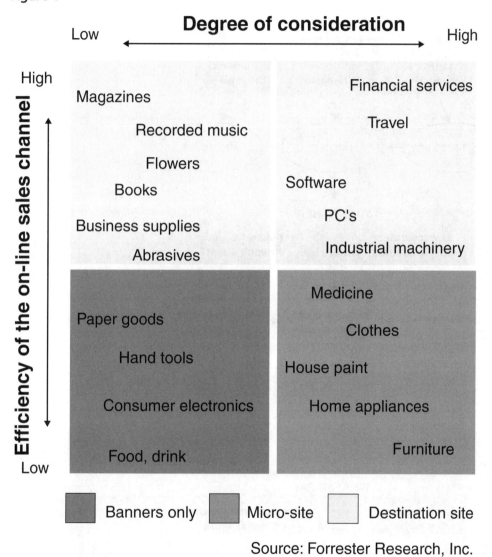

Banners Are Enough for Most Off-the-shelf Products

Most consumer goods companies should use their resources to:

▶ Build interactive banners. Banners should let viewers request free samples, register to win, and order products. Recent examples: HP lets people play "Pong;" First Virtual sells Casio watches; and Metlife provides your ideal weight.

What Advertising Works?

Figure 2

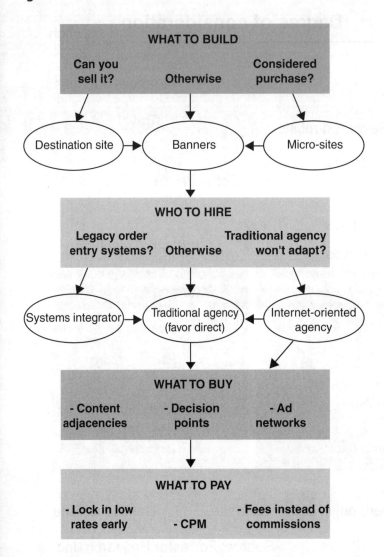

- Sponsor appropriate content. Instead of building a big site, a brand like P&G's Tide should be looking for a way to sponsor the online schedules of every Little League in the country. A logo and positioning line are sufficient to make an impression.
- Maintain a corporate site. Companies need a "face to the public" that catches search engine queries and serves investor relations and recruiting groups. But consumer brands should resist

What Advertising Works?

pumping up the site with product information. Save that money for creative campaigns.

WHAT TO PAY

Ad space and agency pricing mechanisms are still in flux. Smart advertisers will:

- Lock in low rates. Early advertisers can win favorable treatment from grateful sites and networks. Lock up search engine keywords with a right of first refusal. Negotiate a first-sponsor rate in perpetuity.
- Come to terms with CPM. Results-based pricing will appear on the Web - but we don't think that it will dominate. Why? Sold-out sites like CNET don't need to take the risk. Instead advertisers should hold their agencies accountable for click-through results.
- Reset expectations. In traditional media, an advertiser might spend just 20% of its budget with an agency on ad development, placement, and account management -with much of that covered by commissions. The other 80% goes to the media properties. On the web, this ratio is closer to 50:50 for now. So don't expect your agency to survive by commissions - plan to pay fees.

But most of all, plan to use the medium. Most everyone else will be doing likewise.

This report is digested from the March 1997 Forrester Report, "What Advertising Works." In addition to interviews with 51 Web advertisers, Forrester spoke with 35 industry participants from agencies, software companies, and ad networks. These included: Ad Age, AGENCY.COM, Anderson & Lembke, Berkeley Systems, Commonwealth, DoubleClick, FlyCast, Focalink, Foote, Cone & Belding, Grey Interactive, I/PRO, Kirshenbaum Bond & Partners, The Laredo Group, LinkExchange, Modem Media, Narrowline, NetCount, NetGravity, Organic Online, Poppe Tyson, Red Sky Interactive, Riddler.com, Saatchi & Saatchi, SiteSpecific, Strategic Interactive Group, Submit It!, Sunrise Media, TBWA Chiat/Day, The Web Magazine, WebRep, Western International Media, and Ziff-Davis Publishing.

Increasing Advertising Effectiveness on the Web

Title: Increasing Advertising Effectiveness on the Web
Author: Intel Corporation
Abstract: This paper is divided into three parts. Part 1 focuses on the differences between rich, interactive advertising versus banner ads. In part 2 the best ways to target the mass market and overcome delivery issues are explained. In part 3 examples are given from AT&T, Citibank, Delta Airlines and other companies incorporating interactive media technology into their online ad campaigns.

Copyright: ChannelSeven.com

http://www.channelseven.com

Part I

Executive Summary

According to Jupiter, $280 million was spent on Internet advertising in 1996; in 1997, spending more than tripled to $907 million. In an attempt to reach the largest possible audience, many advertisers have limited spending on online advertising to flat banner ads. Recent research shows, however, that rich, interactive Internet advertising is more effective in reaching audiences. Making Internet advertising 'scalable' enables advertisers to deliver the best possible experience to every user-- from the richest, most effective experience for those with high-end systems scaling down to flat banner ads for users with a lower-end system.

Ongoing improvements in PC performance, software capabilities, and bandwidth will continue to offer new opportunities for increasingly rich, interactive advertising on the Internet. And the increasing adoption of powerful Pentium II processor-based personal computers will grow the user base of high end PCs.

Intel supports the expanding Internet Advertising Ecosystem with technical support and vision, along with the tools and research necessary for advertisers to develop rich, interactive Internet advertising.

The Internet as an Advertising Medium

Advertising has always reflected the leading trends in popular culture, and there is no trend more pervasive and compelling

Increasing Advertising Effectiveness on the Web

today than the Internet. The number of Internet users is soaring; the International Data Corporation (IDC) forecasts the number of Internet users to more than triple around the world over the next five years.

Huge Growth Opportunities

The implication of this growth for advertisers is enormous; it means new ways to reach an increasingly large audience with targeted advertising that complements more traditional media and reaches new audiences. Due to rapidly changing technology and more sophisticated ways to communicate over the Internet, advertisers have a new opportunity to develop rich, interactive advertising.The goal of this paper is to explain how Advertisers can take advantage of rich, interactive media to develop more effective advertising for this new medium.

Why Rich Interactive Media Is Better
Consider what makes color television such a compelling advertising medium: It uses pictures, two-dimensional and three-dimensional animation, video, and sound to grab the audience's attention. The Connected PC provides one crucial extra function: interactivity. It allows users to interact with rich content and create a personalized experience. Rich interactive advertising transforms a passive viewer into an active user and one who is more likely to respond to the advertiser's message. Intel uses the term "Rich Interactive Media" to describe visually compelling multimedia (video, 2D and 3D animation,

audio, etc.) combined with interactivity. According to a Grey Interactive/ASI Interactive Research study sponsored by Intel, Softbank Interactive Marketing and the Advertising Research Foundation, the most effective advertising models are those that go beyond standard banners to include interactivity and larger ad sizes.

When an interactive element is added to an online banner, the click-through rate doubles. "This does not mean that banners are dead," said Marianne Foley, senior v-p, Interactive Division of ASI Research, Inc. "It means that for banners to deliver better results, they need to have a playability feature that reaches out to the user and stimulates a response. When you can double your click-through rate, you know the messages you're trying to communicate will have greater recall."

Interstitial Ads Get Noticed

Another major finding of the study is that consumer response and recall to online ads is significantly higher to larger ad units. Interstitial ads, those that "pop-up" on a computer screen as the viewer scrolls from page to page, are the most effective ad models tested. Interstitial ads are noticed, have higher recall and generate higher click-through rates than static or animated banners. The study shows that the recall rate for interstitial ads was 76% and 71% for split screen ads compared to only 51% for banners. The larger ad models also generate a 44% higher click-through rate than banner ads.

Increasing Advertising Effectiveness on the Web

And the good news for online consumer advertisers, according to the Grey/ASI research, is that consumers like the advertising they encounter online. Of the 2400 consumer and business users queried, 56% of the consumers said they liked the advertising. Business users are not as positive about online ads (26%), but interactivity still has positive results in click-throughs. "While business users may be more task oriented in their use of the Internet, our study clearly suggests that a key to attracting click-through is the use of interactive cues," explains Grey Interactive's Norm Lehouillier.

Part II

The Challenge of Varying Viewer Technologies

A key difference between the Web and traditional media is that not all users view a Web site in the same way. Each user has a unique combination of technologies that impacts the ability to access and view more advanced online media. Therefore, targeting the largest possible audience has often meant defaulting to the most common technologies, which typically results in the advertising containing less rich communication, little animation, no 3D, no video or sound.

Three basic elements affect the quality of a user's interactive experience: the type of Web browser used, the Internet connection bandwidth, and the personal computer hardware configuration. There are wide variations in the types of Web browsers used to access on-line content and variations in the plug-in' technologies added to these browsers. Internet connections, which affect the bandwidth, or how much con-tent can be sent from the Web site to the user's Web browser and how quickly it can be transmitted, vary from older 14.4 kilobytes per second (KBPS) modems to 6 megabits per second over cable modems. As for hardware, today's connected PCs with Pentium II Processors allow users to receive and run high-quality video, audio and three-dimensional animation; however, users with older computers may not have the same capabilities to receive and run rich interactive media.

The most common form of online advertising is a simple one-by-six-inch image usually strung across the top border of a Web page. This image is usually in the graphic interchange format (GIF) and because of its size and placement is typically referred to as a "banner". At their simplest, GIF images consist of a single still image, the on-line equivalent of a billboard. The next level accessible to the largest audience includes a simple, two-dimensional animation limited to four or five frames. Neither of these non-rich formats can compete with the emotional engagement and audio-visual entertainment of a good television ad, reinforcing the belief that Internet advertising is boring.

Scalability: The Secret of Mass-Market Customization

But what about those users who don't have the hardware or software to support a rich interactive online experience? The solution is "scalability", which allows advertisers to maximize their audience and deliver the best possible experience to each user based upon which technologies they have. An advertiser develops the rich interactive version first, then creates other versions to match user technologies. By first designing a rich version for users with high-end PCs, advertisers start with the most effective message possible. A good analogy is watching the same advertisement on a black-and-white television compared to a color television with a great sound system. The black-and-white television audience's experience is pretty good, while the color television audience's is immeasurably enhanced. Similarly, the technology of a high-performance connected PC can enrich its users' online experience with audio, video and 3-D animation, while defaulting to a standard GIF banner for users with a low-end platform. An added benefit of scaled Internet advertising is a built-in test of all versions - click-through concept and media selected will affect results, but consistent use of scaled versions will show which rich, interactive technologies are most effective in delivering a strong brand message.

Solutions for Rich Interactive Media

New technologies continue to offer more effective messaging capabilities to overcome the obstacles inherent in reaching an audience with a variety of hardware and software platforms.

While almost all web browsers have the ability to display a GIF banner, advertisers can use all of the following techniques to go beyond the basic GIF format to create rich, interactive advertising solutions.

- Targeting browser plug-in technologies
- Creating advertising in Java
- Using Streaming Media to overcome bandwidth limitations
- Using scalability detection techniques to match richness to capabilities
- Using infrastructure solutions to more efficiently distribute advertising of Rich Interactive Advertising

Plug-In Technologies

One way to enhance a basic GIF banner is through the use of software plug-ins. Plug-ins are Web browser extensions that allow the browser to handle other media types, such as audio, video and 3-D graphics. In order to use a plug-in, the user must download and install it on his or her computer. Plug-ins are written for specific hardware and operating systems, so users must match the right version of the plug-in to their computer. There are hundreds of different plug-ins available on the Internet today, but only a few plug-ins, such as Cosmo and WorldView VRML players, Shockwave, RealAudio, Flash* or Vxtreme, are widely deployed. Generally, the most popular plug-ins are bundled into the latest copy of the browsers. Plug-ins extend browser capability, but their use in supporting rich interactive advertising is somewhat limited. Typically, users will take the trouble to find and install a plug-in to view content but will not bother simply to view an advertisement. Thus, advertisers should target their media buys to sites that promote rich content where users are more likely to have already downloaded the required plug-ins.

Java

An alternative to plug-ins is Java. Java encompasses two capabilities. As a programming language, it allows software developers to write a variety of applications, including those for video, audio, animation and the other prerequisites for rich interactive media. Then there is the Java Virtual Machine, a machine-independent run-time environment that enables any Java application to run on any compatible computer hardware/operating system. The Java Virtual Machine also provides a built-in for downloading Java applications from the Internet. From an advertiser's perspective, Java is important because Java applications can handle rich interactive media files and run on any type of computer that has a Java-capable browser. And whereas plug-ins have to be pre-installed, Java applications can be delivered simultaneously with the content. Java applications require more processing power than applications written for specific hardware platforms, but Pentium II processors easily run Java media applications in a fast and responsive manner.

Streaming Media

The user's bandwidth also affects the amount of data that can be sent from the Web site to the browser. More bandwidth allows the transfer of more data. Greater bandwidth is necessary because richer media typically requires larger files, and most applications stipulate that the entire file be transferred to the user's PC before it can be played. However, on a connected PC with a 28.8 modem, downloading a 30-second video clip can range from several minutes to hours, depending on its quality, and very few viewers will want to wait that long to see an advertisement.

To circumvent bandwidth limitations, a technology called "streaming media" allows the user to view the file as it is downloaded, rather than having to wait until the download is complete. With the latest compression technology, it is possible to use streaming media to download audio, animation and even low-quality video to users with a 28.8 or 56K modem.

Scalability Detection Techniques

Intel has identified three ways to scale rich, interactive Internet advertising: server-side detection, client-side detection and scalable media. Server-side detection involves placing specialized software on the Web server to determine the capabilities of the requesting browser, for instance, to decide whether the browser can run Java applications. The software can also extract information from the Internet address of the requesting user to determine whether the user is connecting through a high-speed cable network or from a dial-up network, such as America Online. Client-side detection is similar to server-side detection, except that the user's capabilities are determined by the user's computer (known as the (Eclient) rather than by software on the Web server.

Client-side detection requires that the user's browser support some type of browser extension, like scripts that can determine whether a specific plug-in is available. Most versions of browsers introduced in the past two years have this capability. Using detection information (both client and server), it is possible to make decisions about what type of content to send to a particular user and to send the user content most suited to the capabilities of their platform. In addition to detection capabilities, some media is by its very nature scalable. For instance, 3D animation can be displayed in Web browsers using plug-ins and a language known as VRML (Virtual Reality Markup Language). VRML sends information about how to build a 3D scene. The browser/plug-in uses this information to construct the scene and render the animation. This type of media scales naturally depending on the performance of the processor. While some machines may render this type of animation as a series of poorly rendered jerky frames, Intel's Pentium II processor renders high fidelity graphics and smooth animation sequences.

Infrastructure Solutions

Until recently, each advertiser had to negotiate the delivery of its online advertising with each web site. This led to inefficient distribution, inconsistent campaign reports, bandwidth inefficiencies, and limited availability of new technologies across sites. In addition, each web site had to manage separate ad servers for individual technologies.

Centralized ad management is an emerging solution that enables the efficient delivery of rich Internet advertising and uniform profiling, tracking, and reporting. Advertisers work with centralized ad managers, such as AdKnowledge, IMGIS and MatchLogic, to serve ads across multiple web sites. Because the centralized ad manager serves the ad directly to the Web user, content is delivered without interruption to the Web user by the Web site. Network enablers, such as InterVU, and Unicast, overcome the difficulties of bandwidth so that rich ads are delivered to Web users in a timely fashion, no matter what the browser or bandwidth capabilities are.

Part III

Putting Rich Interactive Advertising to Work

To demonstrate the potential of rich, interactive, scalable advertising, Intel engaged Modem Media, a leading Interactive Agency with a strong strategic focus on consumer brand building online. Modem Media managed the strategic,

Increasing Advertising Effectiveness on the Web

creative, production, media and research efforts for a collection of leading consumer advertisers - AT&T, Citibank, Delta and Reebok - working with selected technology companies and Web sites to create real-world examples. Their stories, and the technologies used to develop and deliver a scalable experience, are described below. Demos of this work will be available on Intel's web site in January.

AT&T

AT&T's "Drag & Drop Magnetic Poetry Contest" is aimed at college students, new to the choice of long distance and other telecommunications services. Players use their mouse to create poetry from a list of 48 words and then submit their entries - all without leaving the site where the contest was advertised. The winner will receive $1,000. The advertisement will run on Infoseek, Pathfinder, Sportsline, the Double Click Network and RealNetworks Timecast. AT&T used Narrative's Enliven as well as server-side extensions, Java applets and 2-D interactive animation.

Citibank

http://www.citibank.com/
Citibank's banner advertisement of its new US Internet PC Banking product is scalable to end-user computing environments. Consumers with low-end browsers view a banner with the words "Save, Time. Money. Paper. Stamps. Envelopes. Aspirin." scrolling across it. This is immediately followed by "Citibank Direct Access. The #1 PC banking service is free." Consumers with high-end browsers can click to a complete online demonstration of Citibank's Direct Access product, after which they can access Citibank's website, download the software and new customers can use Direct Access to begin the process of applying for a checking or savings account. This rich advertising approach utilizes both scalability and interactivity, key ingredients to effective ads on the web.

Citibank used Narrative's EnLiven, streaming media and a Java applet to send a 2-D interactive banner. A key feature of the banner is that it demonstrates Citibank's new Internet banking service interactively, without requiring users to leave the site they are visiting. An http server-side extension that implements server detection functions determines if the user has a Java-capable browser; if not, the user is sent a GIF banner. The Java animation shows the inherent scalability of Java technology on Intel's Pentium II processor.

Delta Air Lines

Delta Air Lines provides a travel guidebook for each of the cities and regions to which the airline flies. When users click on a three-dimensional image of the planet Earth that's spinning on the advertising banner, a Delta airplane flies past in 3-D space; in its wake, a dialogue box appears from which users create a customized list of Delta-served cities. Another click calls up visitor information on those particular cities. The advertisement will run on The NY Times Online and The Weather Channel.

For Delta, Ligos Technology used V-Realm Builder 2.0 and VRML (Virtual Reality Markup Language) to create a 3-D animated interactive banner advertisement. Client-side detection ensures that a rich version of this advertisement is sent only to viewers who have the software capability to receive it. Instead of sending an URL to a GIF banner, users are sent a JavaScript containing two URLs - one for a GIF banner and one for the VRML banner. The JavaScript looks for a VRML plug-in on the user's browser; if it's not available, the script automatically displays the GIF URL instead. The source code for this script can be found at the Ligos Website at http://www.ligos.com/.

Key features demonstrated include: the use of a client detection script to send different versions of the ad to users with different capabilities and the innate scaling of 3-D animation, which runs best on Pentium II PCs. In addition, the extra cost of producing a GIF banner was negligible because both the GIF and VRML versions were planned from the start.

Reebok

Reebok chose to use rich interactive advertising for a one-time promotion of its new basketball shoe, "The Answer." Aimed at boys and girls aged 12 to 19, the advertising features Philadelphia 76er's guard Allen Iverson and runs on the NBA.com Website through December. After playing the interactive games in the advertisement, users can click to a site offering information about the shoe as well as to a trivia contest with a grand prize to play Allen Iverson in a game of horse at the NBA Jamfest.

Modem Media created several advertisements for the Reebok promotion. "Steal the ball from Allen" is an interactive banner showing a video/animation of Iverson dribbling a basketball while the user hears the ball bouncing. When the user clicks on the banner, the camera zooms in to cover his shoes and the ball. A countdown occurs: "Get ready. 5-4-3-2-1. Go!" and a whistle blows to start the game. Each time the user tries to steal the ball, Iverson performs his signature crossover move to protect the ball until a "Give up?" button is displayed and the user can go to the product information and trivia contest sites. The "Steal the ball from Allen" banner was created using Geo Publishing's Emblaze Creator. Emblaze Creator supports streaming audio, video, and animation without using a browser plug-in.

In "Ball to shoe to Allen," the camera zooms in, then retreats from a bouncing basketball, again with sound effects. Each time the user clicks the mouse, a word appears to form the phrase, "You already know the Answer." The basketball disappears, and is replaced by a shot of Iverson's Answer-shod feet and shins. An "up" arrow prompts the user to pan the camera up Iverson's entire body to get the basketball held above his head. Creator was used to build this interactive banner.

In "Want to play horse with Allen Iverson?" video from the previous night's game streams into the banner space and provides the answer to one of the trivia

Increasing Advertising Effectiveness on the Web

contest questions. Reebok's Director of Interactive Marketing, Marvin Chow is pleased with the response. A teaser ad had 8,000 click-throughs in its first nine hours. "We looked to break out of the traditional banner advertising mold and further position us as a brand of the future. This was a natural fit," Chow says.

Reebok used RealVideo to display a video banner that included information about the new Answer shoe along with a video highlight of basketball player Allen Iverson. Client-side detection searches for a RealNetworks RealPlayer or RealPlayer Plus plug-in. RealPlayer and RealPlayer Plus use streaming technology to provide audio and video displays without the delay associated with file downloads.

Levi Strauss & Co.

http://www.levi.com/
Levi Strauss & Co., while not a part of the Modem Media project, is included in this paper because they use only rich Internet advertising. Levi's chooses not to do scalable advertising as they seek to drive their brand image as a fashion innovator.

Recently, Levi's launched an interactive program called "I-candy". To access I-candy, users click on a gateway icon of a winking eye to download a 15-to-30 second experience that, says Sean Dee, Levi's Media Director of Global Marketing, "makes them think, 'Wow, I hadn't thought of the brand that way before.'"

"Snowflakes" is a 20-second animation of two snowflakes which slowly form and crystallize as a caption spells out, "No two pairs are the same." The snowflakes are fractally generated, so that their images constantly vary, reinforcing the theme that no two pairs of Levi's 501 jeans are alike. "Cowboy" is a game in which users score points by shooting the hat off the authentic Levi's cowboy. In yet another I-candy, a series of shapes appears on the screen and gradually contracts to form the words "Shrink to Fit." As the caption disappears, it's replaced by the slogan "The Original Levi's 501 Jean."

I-candy is targeted to Levi's core consumers, young men between the ages of 15 and 24 who also identify Web use with being hip. "We try to place I-candy in youth-oriented sites where the consumer is coming for an experience, not for utility," Dee says, which is why I-candy shows up on MTV, Addicted to Noise and other music-related sites, as well as SonicNet, Sports Zone and some of the Star Wave sites.

Unlike many online advertisements, there's no forced link to Levi's Web page. "From a quantitative standpoint, one of the benefits of not linking back to the site was that people who had seen our I-candy spent a considerable amount of time longer at levi.com than the average visitor," Dee says. Another benefit of the I-candy campaign is that it's a more efficient way to keep Levi's Web presence fresh than constantly updating the Web site. "I-candy makes it possible to energize the brand," says Dee. "That's why rich advertising is so important to us."

Intel supports this rich-only approach, but realizes other advertisers might find the rich-only approach too limiting. Scalability allows every user to get the best experience based on what technologies they use.

Intel's Interactive Advertising Role

Intel supports the expanding Internet Advertising Ecosystem with technical support and vision, along with the tools and research necessary for advertisers to develop rich, interactive Internet advertising.

Intel will continue to work with Advertisers, general Advertising Agencies, Interactive Ad Agencies, Advertising Networks, Measurement companies, Industry groups, and Publishers to provide leadership into the next generation of rich interactive content technologies.

Industry Groups Summary

With the continuing explosive growth in Internet usage, the market potential for online advertising is enormous. Rich interactive media delivers a more effective message to a target audience than other forms of advertising on the Web.

By bridging the technology gap between users with high-performance and low-performance PCs with techniques such as plug-in technologies, Java, streaming media and scalability detection, advertisers can reach the maximum number of users while matching richness to capabilities for the best-possible impression.

Ongoing improvements in PC performance, software capabilities and bandwidth will continue to offer improved opportunities in this medium and technology adoption will continue to increase the potential audience for rich advertising.

Visit ChannelSeven.com http://www.channelseven.com and subscribe to ChannelSeven.com's free email newsletter!

Justifying the Web for Your Business

Title: Justifying the Web for Your Business
Author: USWeb Corporation
Abstract: The Internet and corporate Intranets are changing the very foundation of private and business life and the World Wide Web is a key factor in this revolution. Establishing a Web site can be one of the most important business moves you make. Because of the enormous potential and significant investment involved, it is essential that you make this move carefully. This paper discusses the importance of thoughtful Web site planning, effective implementation and continuing refinement in tapping the full potential of the Web. It presents a six-step process to ensure that your Web site meets your business objectives, now and in the future.

Copyright: © USWeb Corporation
Biography: see end article

Executive Summary

The Internet and corporate Intranets, particularly the World Wide Web, are changing the way we do business. These networks empower people to get the information they need, quickly and easily, regardless of its physical location. In addition, they provide a high level of interaction between people and information, so the information delivered can be custom-tailored to meet the needs of each individual. As a result, the Internet and Intranets are growing at an unparalleled rate, and bringing about a revolution in business and communication.

Today, approximately 400,000 Web sites are in operation, and more are being added daily—not only on the Internet, but also on private Intranets. In fact, at the beginning of 1996, nearly 25 percent of Fortune 1000 companies had already implemented Web servers on their Intranets, and 40 percent more were planning to establish them.

Implementing a Web site may be one of the most important moves your company makes. Web sites, with their nearly universal reach and highly-interactive nature, present opportunities that are not available through other means. Through Web sites, organizations can increase revenues, decrease costs and build tighter relationships with their customers, employees and business partners.

But Web sites can require substantial investments to create and maintain. Web site spending by companies ranges from $15,000 for small companies to more than $1 million for large companies. Whether you are considering a Web site for the Internet or your Intranet, it's important that you plan and implement it carefully. Only in this way will you realize the full potential of your site and gain a handsome return on your investment.

Planning and building a Web site requires expertise in a wide variety of new areas, including Web technologies, the unique aspects of the Web as a medium and the cyberworld resources available, such as

Justifying the Web for Your Business

Web search engines and Web advertising, to help you generate traffic to your site. To get on the Web quickly, without sacrificing the effectiveness or quality of your Web site, you may want to seek assistance from outside organizations that specialize in Web planning, deployment and refinement.

Background

Web technology makes possible exciting new business models for marketing, communications, commerce, publishing, advertising, client/server applications, telephony, business process optimization, entertainment and even broadcasting. With a Web site, an organization can reach a worldwide audience of literally millions of people, quickly and effectively. Because the Web is interactive, information can be custom tailored for each person for maximum impact. That's why organizations and individuals are implementing Web sites at an astonishing rate.

ActivMedia predicts that Web site expenditures will reach nearly $2.6 billion by 1998. The Yankee Group estimates that annual Web site spending by companies with less than 100 employees ranges from $15,000 to $25,000. For companies with between 100 and 500 employees, that range increases to $75,000 to $125,000.

For companies with more than 500 employees, annual Web site spending is estimated to range between $250,000 and $1 million.

> **Key Statistics**
>
> ▸ There are about 400,000 Web sites in operation today.
> ▸ Web site expenditures will reach nearly $2.6 billion by 1998.
> ▸ 31 percent of today's Web-based businesses are profitable.
> ▸ IDC estimates that there will be 199 million Internet users in 1999.
> ▸ 40 percent of business computer users and 30 percent of home users are already connected to the Internet.

In a June 1996 survey of 1,100 businesses that are already conducting business on their Web sites, ActivMedia found that 31 percent said they were profitable, and 28 percent more said they expect to be profitable in the next 12 to 24 months. These 1,100 businesses accounted for $130 million in Web commerce revenues in June 1996—an annual run rate of over $1.5 billion. This $130 million represents only a small portion of the total business conducted on the Web.

The potential audience on the Internet is enormous. More than 35 million Americans now use the Internet, 9 million of whom started using it in 1996. International Data Corporation (IDC) predicts that in 1999, there will be 199 million Internet users. The Web is a particularly attractive medium because it reaches consumers as well as business users. Today, 40 percent of computer users in business are connected to the Internet compared to 30 percent of home users. IDC expects the percentage of both to more than double by the year 2000, increasing to 95 percent of home

Justifying the Web for Your Business

and 95 percent of business computer users connected.

A business-oriented approach to effective Web sites

Because of the attractive potential of Web technology—such as its worldwide reach and ability to interact with users—and the apparent ease of building sites, many organizations are rushing headlong to establish Internet and Intranet Web sites. But many are taking a haphazard approach, resulting in wasted money and, more significantly, lost opportunity. Forrester Research found that one of the most common mistakes companies make when implementing Web sites is not having a clear vision or purpose for the sites.

Web sites can represent a significant investment in time and resources. That's why, whether you are considering a Web site for the public Internet or your corporate Intranet, it's important that you pursue a well thought-out process in planning, implementing and monitoring your site.

The following sections of this paper present a business-oriented approach to planning, deploying and refining your Web site. By addressing the factors identified in this process, you can take full advantage of the power of Web sites to maintain a competitive edge. There are six fundamental steps in the process:

Step 1: Understand the medium
Step 2: Plan your Web site
Step 3: Deploy your Web site
Step 4: Market your Web site
Step 5: Analyze the results
Step 6: Refine and maintain your Web site

These steps are presented individually in the following sections.

Step 1: Understand the Medium

Before you can realize the full benefits of a Web site, you need to understand the capabilities of the Web and the exciting possibilities unleashed by these capabilities. There are three stages to reaching this understanding: lay the foundation; understand the possibilities of Internet Web sites; and understand the possibilities of Intranet Web sites.

Lay the foundation

In laying the foundation, it is important to consider three factors.

First, understand the nature of the Internet and corporate Intranet, in particular the nature of the Web as a unique and dynamic medium. The Web has aspects that are similar to other media: it can be used to disseminate information, to target specific audiences and to generate direct response. It also has many aspects that are new and different from other media.

One of the most important differences of the Web as a medium is its interactive nature. According to an article in the September 23, 1996 issue of Business

The Ultimate Internet Advertising Guide

Justifying the Web for Your Business

Week, "The successful Web players are not simply replacing existing businesses in the new online medium, they are taking full advantage of the unique, interactive nature of the Internet." This applies equally to Internet and Intranet Web sites. To exploit interaction fully, you need to become aware of the dynamics of users' interaction with Web sites.

The second factor to consider in laying the foundation is to understand your competitors' Web presence. This includes an understanding of the brands and products they are emphasizing, their promotional plans and main messages, their target audience and how they are reaching these targets.

The Web itself provides a rich source of competitive information. Many of your competitors probably post a wealth of information about themselves on the Web. You can use this information to learn about their products and positioning. You can also evaluate their Web expertise by the quality and functionality of their Web sites.

The third factor to consider in laying the foundation is to understand the flexibility and responsiveness made possible by the Web. The Web provides an exciting opportunity to experiment and learn. Unlike more static media, you can quickly incorporate new ideas into your site and observe the effects of changes to content, organization and navigation. According to the Business Week article cited earlier, "The successful Web trailblazers exhibit the ability to adapt, to scrap what's not working and improvise a new business plan on the fly."

Understand the possibilities of Internet Web sites

After you have laid the foundation for learning, you can familiarize yourself with the potential of Internet Web sites. Remember, the Internet extends your reach to a worldwide audience, and it allows you to interact with that audience. The possibilities are exciting:

> **An Internet Web site can help you:**
> - Reach new audiences.
> - Sell products and services.
> - Generate brand awareness.
> - Increase customer satisfaction.
> - Disseminate information.
> - Receive feedback.
> - Automate business processes.

Reach new audiences. The Web provides a new and unique opportunity to reach audiences that have been impossible or hard to reach with traditional media. For example, the 18 to 34 year-old audience represents an estimated 40 percent of Internet Web surfers. Not only can you reach new audiences, you can also present richer messages than with other media. That's because the Web's multimedia capabilities can include audio, video and animation.

Sell products and services. A recent report by Forrester indicates that consumers are eager to buy products online. The report predicts that $518

Justifying the Web for Your Business

million worth of goods will be sold online in 1996 alone, and will increase to $6.6 billion by the year 2000. Many companies use the Web as an additional sales channel that augments their existing, traditional channels. Through the Web, these companies are exposing their products and services to new audiences that might not be available to their traditional channels. Large as well as small businesses are taking advantage of the Web. Dell Computer, for example, opened its Web store in July 1996. The company expects to sell $20 to $30 million worth of PCs per quarter from that site.

Enhance brand awareness. A Web site can deliver a richer brand identity than other media. You can augment traditional text and graphics with more engaging multimedia, including animation, audio and video. More importantly, Web technology helps you develop more individualized relationships with your customers by enabling you to deliver information that is custom-tailored to each customer. As a result, you can generate brand awareness that has increased personal meaning to each customer.

Increase customer satisfaction. Because of its extensive reach and high level of interaction, a Web site can help you provide better service, better information, better support, and develop a closer relationship with your customers.

Disseminate information. Through your Internet Web site you can provide easy access to information about your products, services and company. This helps you move customers more quickly to the next step in the sales cycle.

Receive feedback. Solicit feedback from Web site visitors. Customers are more apt to respond because it's so easy and anonymous. They simply enter the information requested on a questionnaire or survey page and click a button to send it to you.

Automate business processes. Automate a variety of business processes by redeploying them to a Web site. Federal Express, for example, permits customers to check the status of their packages on a Web site. As a result, up-to-date information is readily available to customers without the cost of additional support personnel.

Understand the possibilities of Intranet Web sites

Intranet Web sites can plug into your existing network infrastructure. As a result, they leverage your network investment. Because they operate over your existing network, Intranet Web sites are easier to secure than those on the Internet.

An Intranet Web site can help you:

- Automate business processes.
- Redeploy client/server solutions.
- Disseminate information.
- Facilitate a collaborative culture.
- Increase employee satisfaction.
- Receive feedback.
- Automate business processes.

The Ultimate Internet Advertising Guide

Justifying the Web for Your Business

Like Web sites on the Internet, Intranet Web sites present a number of possibilities that can revolutionize the way you do business:

Automate business processes. Automate a variety of internal business processes on your Web site for increased efficiency. For example, many organizations are using Intranet Web sites to automate the distribution and administration of internal documents, including policies and procedures, benefit selections, financial information, telephone lists and job postings. Electronic distribution eliminates the high cost of updating and distributing paper documents every time an update occurs.

Redeploy client/server solutions. By redeploying client/server solutions to Web sites, you provide universal access to information without the need to install and manage specialized client software. Users can access the information they need through their standard Web browsers. Many organizations are already redeploying client/server applications in human resources, accounting, sales management and executive information services onto their Intranet Web sites. These Web-based applications are considerably less expensive to maintain and manage than traditional client/server solutions.

Disseminate information. Through your Intranet Web site you can disseminate internal information to employees—and even to business partners, such as suppliers and contractors. You can use your Web site to present information in new and engaging ways, so your employees will be more apt to access the information. Your business partners will also appreciate being kept "in the loop."

Facilitate a collaborative culture. Because important information flows more freely through Intranet Web sites, it is easier for your employees to become more engaged, involved and interactive—within their own departments and workgroups as well as with other departments and workgroups.

Increase employee satisfaction. Use your Web site to keep employees informed and solicit their feedback on matters that are important to them. As a result, you'll build closer relationships with them.

Receive feedback. Because of its ease of interaction, a Web site can help you obtain valuable feedback from your employees and business partners. Use this information to improve service and support to these people who are so important to your business.

Step 2: Plan your Web site

After you have developed an understanding of the Internet and Intranet, you are ready for the next step—planning your Web site. This is, perhaps, the most important step because it establishes how your Web site fits into your business objectives.

Justifying the Web for Your Business

> **Important issues to consider in planning:**
>
> - Plan your site within the context of your overall business strategy.
> - Define your goals.
> - Identify opportunities to increase revenue.
> - Identify opportunities to decrease costs.
> - Define your target audience.
> - Establish and monitor objectives.

Plan your site within the context of your overall business strategy.
In the planning step, you determine the opportunities the Web presents to your organization and then define your objectives accordingly. It is important to develop your Web site business plan within the context of your overall business strategy. Be sure to address two major issues:

How does the Web site support your existing business objectives? Does the Web present new opportunities that are not currently available to you? These issues / ideas should be integrated into your overall Web business plan. Look for opportunities to increase revenue and decrease costs. In defining your objectives, it is important to determine your target audience, and to build in the means to monitor your progress in accomplishing your objectives.

Define your goals
Once you have an understanding of how the Web fits your business, you can determine your goals for establishing a Web presence. Only in this way can you tap the synergy between the Web and other marketing tools available to you. In particular, you need to establish the role of your Web site in your overall marketing mix. Businesses often consider the following opportunities when developing their Web site:

- Strengthen brand awareness
- Enhance product awareness
- Increase sales
- Improve communications with employees and customers
- Automate business processes
- Reduce overhead sales and printing costs
- Stay ahead of the competition

Identify opportunities to increase revenue
The Web presents a variety of ways to increase revenue:

Boost lead generation. Take advantage of the interactive nature of the Web to capture leads.
Speed lead response. Because you get leads immediately, you can respond faster, while the customer is still "hot."
Reach new customers. Because of its worldwide reach, the Web helps you reach customers who may not be available through other media.
Add a new sales channel. Add online sales that are incremental to those of your traditional off-line channels such as retail and direct mail.
Increase sales through existing channels. In addition to providing a new sales channel, the Web can also help boost sales through your existing channels by increasing product and brand awareness.

Justifying the Web for Your Business

Improved customer service and support. The Web presents opportunities to increase revenue indirectly through enhanced customer service and support.

Identify opportunities to decrease costs
You can use both Internet and Intranet Web sites to reduce costs and improve productivity in external and internal business processes:

Reduce support costs. It is often cheaper, easier and more effective to support customers over the Internet than through more traditional methods such as telephone support.

Reduce sales costs. Sales over the Internet typically require less overhead and less sales support than traditional sales channels. A Web site can reduce dependence on more expensive sales channels, including retail.

Reduce inventory costs. A Web site can help you reduce inventory costs by shortening sales cycles.

Reduce materials costs. Save paper production, printing and distribution costs by disseminating information electronically over the Internet or Intranet. For example, you can publish annual reports, distribute marketing materials and present customer support tips on your Internet Web site. An Intranet Web site can lower the cost of delivering internal manuals and forms.

Define your target audience and their motivations
The Web gives you the ability to reach audiences that are not accessible using other media. But reaching your audience is only part of the task. It is essential that your Web site be carefully tailored to your target audience. Just as with any other medium, you need to know your audience to take full advantage of the Web. Ask yourself the following questions about your target audience:

Who are they. Are they consumers, business customers, business partners, government organizations, or your own employees? What are their demographics?

What are their qualifications? What is their level of familiarity with computers? Are they "net-heads?" Are they technically oriented? Are they consumers?

How will they access your Web site? Will they use network or dial-up connections? What is their typical modem speed? What kind of browsers will they use? These factors influence the content of your site. For example, if your target audience typically uses dial-up connections, you should not include graphics that will require long download times.

Establish and monitor objectives
You need to set up methods and metrics to monitor and evaluate your Web site—on a continuing basis. The site can provide feedback in two ways: It can collect data automatically, such as the number of visits to each page and the paths visitors take through the site. You can also solicit feedback from visitors through questionnaires and survey forms on your site.

The metrics you establish for monitoring should be objective and measurable. Use them to evaluate the effectiveness of your

Justifying the Web for Your Business

site in meeting the objectives you have established. Are you reaching your intended audience? What incremental sales are attributable to the site? What cost savings have you realized by automating business processes on a Web site? Are you finding increased employee satisfaction because of improved information flow through your Intranet site? Is your overhead reduced because of lower printing costs (external or internal) or lower sales costs?

Step 3: Deploy your Web site

After you have completed the understanding and planning steps, it's time to deploy your site. In deploying your site, take into account the following factors:

Design your site from a usability perspective and not merely from an aesthetic perspective. Rather than just looking for an award-winning, "cool" site, consider the site visitors. Make the site easy to navigate. And don't scare visitors away or frustrate them with a multitude of snazzy graphics that take forever to load. **Build in the means to monitor and continually evaluate your Web site.** Treat your site just as you would any other business tool.
Take advantage of the highly interactive nature of the Web. Solicit visitor information.

What you'll need
There are a staggering number of Web-related tools for creating and posting Web sites, and new tools are appearing every day. The cost of evaluating, purchasing, integrating, learning and using these tools can be significant. And it often requires expertise in new scripting languages such as HTML. In addition, if you want to add pizzazz and include such capabilities as soliciting feedback, you'll need expertise in other, more complex programming languages such as Java, ActiveX, Shockwave and CGI Scripting.

Your site should be easy to navigate, so that visitors can quickly move to the pages they need. (How many times have you left a Web site in frustration because you couldn't easily navigate it or get the information you want?) A well-designed site requires expertise in the unique interactive aspects of the Web medium. Fortunately, there are organizations that specialize in Web site planning, design, construction, deployment, monitoring and maintenance. These Web professionals can provide you with the expertise you need to create and deploy an effective Web site.

Many organizations are already turning to Web professionals for assistance. IDC and Network World jointly conduct an annual survey called the Network World 500. In this survey, they interview 500 organizations that are among the leading networking users in the United States. These are companies with more than 1,000 employees, multiple sites with internetworked LANs and WANs and annual networking expenditures of more than $5 million. The 1996 survey reveals that 23.1 percent of these companies are using outside firms for installation and maintenance of their Web server hardware and software, and 22.1 percent are using outside services for their Web site content

The Ultimate Internet Advertising Guide

Justifying the Web for Your Business

design and maintenance. Smaller companies, with 100 to 1000 employees, are outsourcing at an even greater rate.

Integrate your objectives
Before actually building your site, you should determine how your site can help you accomplish your business goals and objectives. Take a holistic approach, that is, don't look at the Web site in isolation but rather in the context of your overall business plan. Understand how your cyberworld objectives relate to your physical world objectives. Understand the role of each medium within your marketing mix, and leverage the strengths of each.

Try to make it all work together. Support each marketing objective across multiple media where possible. For example, cross-promote among media. Strive for synergy and consistency. Even though the individual messages may vary to leverage the unique strengths of each medium, the overall flavor should be consistent across media.

Build your site
After you have integrated your objectives, it's time to build your site. Using the three-phased approach outlined below will help you get on the Web quickly without jeopardizing quality or effectiveness. It will also provide a valuable learning environment, enabling you to adjust your business model to get maximum leverage from your Web site as you gain experience.

First, repurpose existing materials to the Web site in a compelling manner. This does not mean merely copying existing paper documents to the Web server. It means adapting existing materials for online use.

Adding a table of contents that provides hypertext links to document sections facilitates navigation considerably. Adding links within and across documents further simplifies navigation. Keep it simple and intuitive for the visitors coming to the site.

Second, begin moving appropriate business processes to your Web sites. Consider both external and internal processes for deployment on the Internet and Intranet. Increase the geographical impact of your Web site. Take advantage of the Web's potential to engage target audiences. You may even begin conducting basic commerce operations over the Internet and Intranet during this phase. Some processes may require that you integrate the Internet with your Intranet. For example, it may not be practical to give an offshore supplier a direct connection to your Intranet Web site. You can however, allow that supplier to connect to your Intranet Web site through the Internet. As you increase the number of processes you deploy, the more your customers, vendors and employees will view your Web site as a viable place to conduct business. The result … you'll improve access and service to your customers, suppliers and employees, and tighten your relationships with them. And you'll increase revenues and reduce costs at the same time.

Third, grow and evolve your Web site. Create active content that can be customized for each visitor. Evolve your site into a full business entity—with its own goals, objectives and strategies. Above all, keep your site dynamic. Your site will become a strong adjunct to many of your organization's business activities and will

Justifying the Web for Your Business

soon be considered as mission-critical by the various groups with which you interact.

Step 4: Market your Website

Merely implementing a Web site, no matter how "cool," does not ensure that your intended audience will visit it. Even when they do visit it, it does not ensure that they will return. As a result of the large number of Web sites already on the Internet, your site can get lost in the shuffle unless it is properly promoted as an extension to your current marketing efforts or business processes. Even with internal Intranets, promotion is essential to ensure that your target audience is aware of them and uses them to their fullest capability.

You need to engage in active marketing to draw an audience to your site and keep them coming back. There are a variety of ways to market and promote your Web site, many using the Web itself.

Leverage existing marketing resources

You can generate Web traffic by extending your traditional marketing programs. This may be as simple as adding your Web site URL to your existing marketing collateral, press releases, advertisements, on-hold messages and product packaging. Once you've caught a customer's eye through traditional marketing activities, let them know that more extensive information is available on your Web site. That helps you build qualified site visitors who are interested in your content. Also include areas that tie into your latest non-Web marketing activities and promotions.

List with online information and directory resources

Several types of online resources help Net surfers find information on topics of interest, including companies that provide the products and services they are looking for. You should list your site with as many appropriate resources as possible. Unfortunately, thousands of resources are available, making comprehensive coverage difficult. Online resources fall into four categories:

Search engines. These sites utilize indexing software agents, often called robots or spiders, that continually "crawl" the Web. They visit virtually every site in search of new or updated pages. When an agent visits a site, it records the full text of every page and visits all external links. Agents revisit sites periodically to refresh their databases.

Directories. Unlike search engines, directories do not employ indexing software agents. Instead they require you to register your Web site with them. Directories are usually subdivided into categories, so you need to submit your site under appropriate headings. Listing your site in as many relevant directories as possible helps to ensure that visitors find your site when they are searching.

Announcement sites. The explosion of sites being added to the Web has resulted in the establishment of announcement sites that track all new Web sites. These sites announce different types of new Internet documents—such as new Web pages, new articles and new resources. Announcements are posted for only a short period of time. When they remove a document from their "What's New" section, however, most announcement sites archive it so users can continue to access it.

Award site and cool site guides. These guides are becoming a popular source for finding interesting and useful Web sites. Guides post and rate only a small percentage of the sites submitted to them, and typically select only one new site per day. You should ensure that the people who maintain award and cool site lists are aware of your site. Make sure your site is "rate-worthy" before submitting it for rating. Being selected as an award site or cool site attracts high traffic—but only for a short time.

Advertise on other Web sites

Web advertising is maturing as a marketing strategy and is now being brokered by existing and new agencies. Advertising on targeted Web sites can help you build traffic for your Web site.

As with other media, message, location and timing are critical components of a successful banner advertising campaign. Targeting the right audience with the right message provides a greater return than a high number of impressions to an untargeted audience.

Promote your Intranet Web site internally

Your Intranet Web site will be effective only if your employees and business partners are aware of it and use it. That's why you need to promote your Intranet Web site internally. There are a number of methods you can use to promote your site, including:

- Feature the Web site in employee newsletters.
- Distribute flyers or posters promoting the site.
- Post site information on company bulletin boards.
- Encourage managers to talk up the site to their employees and at company meetings.
- Conduct promotional programs such as internal contests based on some aspect of Web site use.
- Conduct training sessions on using the site.
- Keep your business partners informed about your Web site.

Get help from outside sources

As in building a Web site, marketing and promoting a site requires specialized expertise.

Fortunately, there are a number of firms available to help you promote and market your site. These firms offer a variety of services, including:

- Arrangement of strategic links with related Web sites.
- Press release submission.
- Competitive research and evaluation.
- Interactive, banner type Web advertisements.
- Web advertisement placement strategy.

Some firms offer specialized services. For example, there are Web site promotion firms that employ automated tools to register Web sites electronically with search engines and directories.

Justifying the Web for Your Business

These firms typically offer a free service to submit a site's URL to the top general search engines and simple directories. They also offer extended tools that register a site at a considerably larger number of resources, and they can target specific topics or geographic locations. Other firms offer Web site marketing and consulting packages that include a combination of services.

Step 5: Analyze the Results

It is important to monitor and evaluate your Web site continually to ensure that it is meeting objectives. In the planning stage, you determined what data you would gather and what metrics you would employ to analyze that data. Based on the data and metrics, you can analyze your Web site on a variety of dimensions depending on your needs.

There are two primary purposes for analysis. One is to evaluate the effectiveness of your site and compare it to your traditional marketing programs and business processes. The second purpose is to provide information to help you continually evolve and fine tune your site. Remember to evaluate Web results in the overall context of your business strategies to determine if the results meet your business objectives.

In analyzing your site, look at the data you gather online, including:

Number of site visits and specific page visits per day. How large an audience are you reaching? Which pages are most popular? Which pages are least popular?
Visitor paths through the site. How do users typically travel through your site? Which paths are most often taken?
Solicited online feedback. Look at information gathered from solicited online responses, such as questionnaires, that are posted on your Web site. This valuable information can help you optimize your site to the needs of your customers.
Visits derived from ad banners. Analyze how well banners are working. Determine which banners are producing the most visits, and which are producing the most qualified visitors.

Also, look at other data, such as:

Solicited offline feedback: Analyze feedback from sources other than your Web site. For example, look at customers' responses to questions asked by offline sales personnel, such as "Where did you hear about our product or company?"
Testimonial and anecdotal information. Take advantage of unsolicited information as well as solicited information.
Incremental revenue. Look for added revenue that is directly or indirectly attributable to your Web site.
Cost savings.
Look for cost savings that are directly or indirectly attributable to your Web site.

Use the information you gather to compare the results of your Web site programs to those of your non-Web programs. You also need to compare the quality of the results for each of these types of programs.

Step 6: Redefine and Maintain your Web site.

The Web is the most dynamic medium available today. As a result, it is important to establish a philosophy of changing and evolving your site continually to keep it fresh, at maximum effectiveness and in tune with your overall business strategies.

You should establish a process that enables you to refine and update your Web site on a continuing basis. The process should include the following components:

Leverage interaction to improve your site. You will continue to get feedback on your site. Use this feedback to help you modify your site. Establish an iterative process to keep your Web site optimized. Take advantage of the dynamic and interactive nature of the Web to provide a learning experience for your organization.
Experiment and monitor the results. Make changes and watch their effect. An iterative process enables you to involve visitors in bringing your site up to its optimum potential and keeping it that way.
Review site marketing strategies. Evaluate results in the context of your marketing strategies and determine if the results are on- or off-track. Do this for both Web site and non-Web site tactics.
Review business strategies. Reconcile your Web site strategy in the context of your overall business strategy. If you find the site is not meeting certain business objectives, re-evaluate your Web business plan in light of these shortcomings. Establish a process for continually refining your Web site objectives in light of new information. The process should also include updating the role your Web site plays in your overall business plan.
Stay in tune with technology. As technology advances, you need to evolve your Web site when appropriate in the context of your objectives.

Conclusions

A Web site can revolutionize the way you do business. It can help you increase revenues, decrease costs and build tighter relationships with customers, employees and business partners. As you gain experience with your Web site, you can continue to integrate it with your business, taking advantage of its unparalleled reach and high level of interaction to increase your competitive edge.

Establishing and maintaining a Web site can represent a significant investment. To realize maximum return on that investment, you'll need to plan your site carefully, design it to take full advantage of the Web's unique capabilities, market it effectively and refine it continually. This requires expertise in a variety of areas, many of which are new to your organization.

Fortunately, there are a number of firms specializing in helping organizations like yours to plan, create, deploy, market and evolve Web sites. With the help of these firms, you can get on the Web not only quickly, but also effectively. You'll realize immediate benefits, and gain strategic advantage. By establishing a Web site now, you'll position your organization to tap the

Justifying the Web for Your Business

rapidly expanding potential of one of the most exciting mediums to appear on the business scene.

Biography: USWeb is the first professional services firm to provide companies with a single source for successful Internet and Intranet solutions.

USWeb, Outfitters to the World Wide Web, enables businesses to reduce the complexity and cost associated with building a presence on the Web. USWeb's strategy is to offer best-in-class professional services from which businesses of all sizes can obtain high-quality, consistent internet and intranet Web site solutions, including needs analysis, consulting, development, hosting, site marketing, maintenance, and education.

USWeb sets the standard in Web site design, development, marketing, and maintenance. Our experts know Internet and Intranet technology inside and out—so customers enjoy the advantages of unparalleled quality, reliability and security in a Web solution they can easily justify from a cost perspective. USWeb is one of the most experienced developers of high-end Web sites, and unlike other Web presence providers, USWeb offers the strength and experience of a national organization with over 30 offices across the country.

The company has also established strategic alliances with Compaq, Sun Microsystems and other leading Internet technology and communications companies to provide a combination of Web expertise that is unparalleled in the industry.

To find out how USWeb can help you bring your business to the Internet/Intranet, telephone 1-888-USWEB-411, or visit www.usweb.com to locate the USWeb office nearest you.

Collection of DrNunley.com Marketing Articles

Title: Collection of DrNunley.com Marketing Articles
Author: Dr. Kevin Nunley
Abstract: This collection contains a series of articles written by Dr. Kevin Nunley. The articles give powerful marketing and advertising advice and strategies.

Copyright: Dr. Nunley
Biography: Kevin Nunley provides marketing advice and copy writing for businesses and organizations. Read all his money-saving marketing tips at http://DrNunley.com/. Reach him at kevin@drnunley.com or (801)253-4536.

Index Articles:

- Why the Internet Will Stay Strong...even if the economy goes bad. And how YOU can profit from coming bad times.
- How to Work With Charities to promote your business and help your community.
- How to Improve Your Ranking On Search Engines. New techniques can drive thousands of visitors to your web site.
- Search Engines Are Getting Smarter. How the busy business person can draw lots of visitors to their site. Easy step-by-step strategy.
- If at First You Don't Succeed. Five ways to fix slow business and give yourself a BIG second chance.
- Get Listed On Search Engines By Going Through The Back Door Why your site isn't getting listed and what you can do about it NOW. Get all the basics...and some crafty insider strategies, too.
- How to Build Your Business Empire on the Internet for Next to Nothing! This opportunity may not be around forever. Take advantage of this virtually FREE mass media to build your biz and spread your name around the world.
- How to Get People to Trust Your On-Line Marketing! Recent surveys show that many people don't buy things on the 'Net because they don't yet trust this new form of media. Here's what you can do to reverse that trend and energize your on-line marketing.
- Four Part Series: THE INTERNET: Where Future Billions Will Be Made. Plug Yourself In! How to find your market and make your mark. PROMOTE ON THE NET: A big list of ways YOU can get low cost and FREE marketing on-line. GET A WEB SITE THAT SELLS! Ways to make sure your site turns visitors into customers. HERE COMES BUSINESS... Can you handle it? Kevin gives you efficient ways to manage lots of customer questions, orders, and e-mail. Plus a list of places to get FREE promotional tools!

WHY INTERNET BUSINESS WILL STAY STRONG EVEN IF THE ECONOMY GOES BAD.

And what you can do to make sure you come out on top.

Day after day news media warns us the world's economy may be in serious trouble. Storm clouds are on our economic horizon. Japan, once the most profitable nation in the world, is having deep money troubles. Experts say Japan is now where the United States was at the start of our Great Depression back in the 1930s. Russia, Latin America, and Southeast Asia are also having serious economic problems. Hunker down, a world depression could be on the way.

Does this gloom and doom apply to the Internet? Will thousands of small Internet business be forced to close down? I don't think so. Here's why.

Internet business is still brand new. Even the old-timers have only been on-line for three or four years. In many ways, we're just now figuring out how Internet business works. And guess what? It appears to be very different from regular business.

Small and versatile is a big advantage.
Big businesses dominate the traditional business world. The Walmart's and MicroSoft's have steadily forced smaller, family-owned businesses out of the way. Not so on the Internet. Three out of four Internet businesses are very small, often only one person working from home. Some of the most successful web sites are run by a single person still working a regular job. They take care of the business before work, during lunch, and late into the evening.

Small businesses are versatile. They can change directions at a moment's notice. That's a big advantage when times are hard. A big company has specialized employees and materials stockpiled to fill a particular need. If the economy changes and that need dries up, the big biz is stuck. Meanwhile, the one-person Internet business can change its direction in an afternoon. You can take down your big web site offering investment advice and put up an equally impressive web site showing people how to get out of debt. No employees to retrain. No leases to get out of.

Internet business can personalize. Every indicator of how the future will be points to a much greater demand for personalized services. Instead of buying a one-size-fits-all service from suppliers, you will enjoy services and products that are closely tailored to exactly what you want and need. Internet leaders, including Bill Gates, have said they believe the future of the Internet lies in personalized services supplied by small companies and individuals.

The Internet may be at odds with the Market. This idea is a tad complicated, but I think it's important to understanding why the Internet probably won't feel the pinch of a bad economy. Market economics, the basic principles that govern business, doesn't seem to fit the Internet. Market economics generally encourage big

Collection of DrNunley.com Marketing Articles

companies to get bigger, buying up and out-maneuvering smaller companies. The biggest companies dominate their industry. Sometimes they grab a huge percentage of all sales in their particular field. This is very hard to do on the Internet. It may be impossible to build an Internet-based monopoly. I may raise millions of dollars and create the biggest, coolest web site business in history. That doesn't keep you and 1,000 other aggressive folks from doing the same thing tomorrow and taking my advantage away.

What can you do to profit from coming hard times? Economic downturns can be scary times. It's hard to know if you should start or expand a business or keep your money in the bank. Don't spend money you don't have to. Yet economic hard times can pose a terrific opportunity for people working in a new area like the Internet.

While traditional business models stall, Internet business surges ahead on the shoulders of a very different way of doing business.

Make your Internet presence BIG. Expand your web site. Jazz up the look. Add lots of helpful articles, add links to useful sites, and create alliances with other entrepreneurs. Keep your web site as focused as you can. Let people know you specialize in an area or line of products. When customers need a particular thing, they'll know you're the specialist that can give them personalized help.

Finally, remember the wise old saying: When business is bad, advertise. The Internet shows little honor to those who come in with lots of start-up money. Instead, the Net rewards those who are popular. The more visitors your web site and email box have, the more power you have on the Internet. Publicize your web site, your business, and your name. Distill your name and main benefits down to a short, easy sentence and put that sentence everywhere you can without spamming. Advertise in email newsletters. Put banners on sites like your own. Send out press releases to media. Participate in newsgroups.

Paint your promotional efforts with big broad strokes. Spend as much as half your time promoting. By looking big and providing tightly focused products and services to a well defined group of customers, you can ride the Internet wave into the future. It may well be a future that gives the Internet new and greater prominence.

Collection of DrNunley.com Marketing Articles

HOW TO WORK WITH CHARITIES TO PROMOTE YOUR BUSINESS AND HELP YOUR COMMUNITY.

Looking for a smart way to promote your business? How about a tried-and-true marketing method that makes you look good in the eyes of thousands of new customers AND helps people in your community? Notice how many successful businesses in your town support charities. It costs less than you think to help out a non-profit group and the promotional payoff can be huge.

Plus, there are lots of worthy projects in your area that could really use a helping hand. People give more to charities from November to December than any other time of year.

Organizations benefitting children and the hungry get special attention from the public. However, in surveys the public always says they wished fundraisers would be held at other times of year. They are suspicious of fundraising being grouped around Christmas.

GETTING STARTED
NOW is the time to start talking with a good non-profit. Contact their director and ask how your business can help. In most cases you can dedicate a small percentage of your sales to the charity.

Non-profits will be ready and willing to include your name in their advertising and marketing campaigns. Have a logo, flier, short ad copy, or web site banner for them to use. Make yourself available to join charity representatives on radio/ tv talk shows and Internet chats.

There are so many worthy organizations that it is often hard to chose which one to help. If you don't already have a favorite, pick one that relates to your business in some way. If you sell children's clothing, working with a charity that helps underprivileged kids at Christmas would be a good match. Some charities are better equipped to work with sponsors and the media. Others are new, have inexperienced staff, and may appreciate your business experience in showing them how to organize people and resources.

THE BENEFITS TO YOU
Most people don't buy the item with the lowest price. Customers highly value service and image. By involving your business with a non-profit doing important work, you get the notice and respect of thousands of people who otherwise might not know about you. Note how many major newspapers and television stations are promoting charities this time of year. Their audience and advertisers appreciate when media works to improve the community. Your customers and prospects will feel the same about you.

As an added bonus, business, political, and community leaders are often heavily involved with charities. The people you meet can form a valuable network of contacts for future projects and business.

CHARITIES ON-LINE
The Internet started as a non-profit effort and still carries a strong feeling of people selflessly working to improve life. Web

Collection of DrNunley.com Marketing Articles

designer Lisa Schmeckpeper recently found the Net a perfect place to do non-profit work. "It's turned out to be very effective. In working with Toys Not Tears, we've linked our non-profit site to the web sites of participating merchants." Order forms can be modified so when a customer buys, a percentage of the sale is collected by the charity. It's easy and everyone involved benefits from the constant flow of customers from site to site. The group uses one site for consumers with another to recruit merchants.

DON'T BE TOO COMMERCIAL
It's easy to get carried away trying to promote the sponsor's interests in a non-profit campaign. If it appears that sponsors are being promoted more than the work of the charity, the whole thing can backfire. Sponsors who stay discreetly in the background receive more benefit in the end.

Focus on how you can help make things easy for consumers. Ease of participation is often what separates success from failure. "Try to solve the problems a potential donor may have such as no time to write a check and mail it in, no extra money available, and fear their donation may not reach the right people," Schmeckpeper points out.

PROMOTE HARD
Lots of worthy non-profits are shouting their messages this time of year. Even though you are a sponsoring business, you may find yourself helping out on the publicity end. Use every available marketing and publicity option. It takes lots of repetition to have an impact. A well-written press release will interest editors and producers. Many email newsletters are good about donating no-cost ad space for charities. Radio, TV, and newspapers will often give you free time and space if you have a cause or event their audiences will be interested in.

Also think how you might be able to continue your association with a charity year after year. Those who don't notice you this year will be twice as aware the second time you participate. Many of the most successful business-charity associations have been going on for decades.

There's no question your business helps others by providing valuable products, services, and ideas. You'll multiply the good feeling when you lend a hand to a non-profit charity.

HOW TO IMPROVE YOUR RANKING ON SEARCH ENGINES

New techniques can drive thousands of visitors to your web site.

Randy is the first to tell you driving traffic to your web site is an expensive proposition. "I spent the promotion budget on ads, the grocery money on banners, and we still aren't getting that many hits," he says.

Now Randy turns to search engines. He's heard that tens of millions use them each day to navigate around the Net. A good listing on a major search engine can bring thousands of already-targeted customers to his site. But how can he get his site listed-- and fast?

As search engines get more sophisticated, it's easier to get a good site listed. It's also harder to "trick" search engines into listing you. The good news is, if you have a site that is focused on just a few topics your best prospects are likely to search for, you're in a situation where the mammoth Internet portals can help you. And it's all free!

Forget about those services that promise to list you on thousands of search engines. There are only three search engines that really count. Because a number of the majors use the same database, you can register on one and show up on several of them.

"Don't waste time trying to get on Yahoo," says Jerry West, search engine expert with NetGateway. "Instead, submit your site to HotBot.com." Both use the Inktomi database. A listing on HotBot virtually guarantees a showing on Yahoo.

Recently AltaVista surged in user popularity. It ranks as one of the best places to have your site listed. Alta Vista rotates visitors through four databases. If you don't see your site listed, try again in 15 minutes when the next database will be on display. Register with Alta Vista weekly to get your site listed in all four databases.

You will also want to submit your web address to Excite, perennially one of the most used engines. Like HotBot and Alta Vista, Excite spiders through your site in a matter of seconds. Your listing can show up within a few days.

YOUR TITLE IS TOPS.
Search engines pay close attention to the title of your site. That's the line that appears in the little box on your browser. Take care to include two or three keywords that your best prospects are likely to type into a search engine when they are looking for a site like yours.

"Welcome to Sheila's Web Site" may sound good, but it doesn't give search engines much to go on. It would be better to title Sheila's opening page with some words about what Sheila offers, like "Solutions for credit card debt, loan restructuring, and bankruptcy."

That gives a search engine several good keywords to use when categorizing Sheila's site, words that her customers are likely to use when looking for her.

Collection of DrNunley.com Marketing Articles

HOW TO FIND POPULAR KEYWORDS.
Big companies now have access to records from all the major search engines. These databases show which keywords people search with. Often a company can find a keyword that millions of people search with, but isn't used by many web sites. Using a popular but neglected keyword can draw a goldmine of new visitors.

West says you don't have to lay out big bucks for the insider's list of keywords. Anyone can find them by checking Goto.com. The engine lists keywords that people use along with a rating of their popularity.

More and more, getting a good listing on search engines has less to do with tricks of the trade and more to do with how well your site covers a particular topic. After all, that's what people hope to find when they use a search engine.

Spend time crafting a inviting description of your site to include in your Meta Tags (for a quick lesson on how to write your own Meta Tags, see Alta Vista's short tutorial at <http://www.altavista.com> (look in the "add url" section").

Search engines display your description next to the link to your site. A good description can grab attention and cause lots of people to click to your site rather than another one.

Be patient and don't get too obsessive. I find that once sites are listed on the six or seven major search engines, they tend to start showing up on many other search engines and link libraries. You can register your site with all the majors with one click at <http://www.all4one.com/> . You can register free with over 400 search engines and link libraries with a single click at <http://www.submit4free.com/>.

Finally, make sure the copy on your web site is interesting and makes customers want to buy. All the traffic in the world won't earn you a dime unless your web site converts visitors into buyers.

INTERNET SEARCH ENGINES ARE GETTING SMARTER.

How the busy business person can drive lots of visitors to their web site.

Have you had this experience? You go to your favorite search engine, type in a keyword for the kind of web sites you're trying to find, and the search engines comes back with "There are 20,132 pages that contain this information."

Yikes! Where do you start? The people who design search engines have heard your complaints. Most have been working hard to make search engines smarter. Here is how engines are changing and how you can take advantage of these evolving features.

With the exception of Yahoo, which uses real people to review web sites (and, technically, isn't really a search engine), all search engines are computers. When you register your URL (web site address), the computer runs over, takes a quick look through your site, and reports the information back to the search engine's data banks. In general, computers aren't as smart as people, so savvy web designers have come up with all sorts of tricks to talk search engine computers into giving them a high listing.

No doubt you've clicked over to the top two or three listed sites, only to find that they have little to do with the topic you're searching for. That's exactly what search engine designers are trying to get away from.

Increasingly, today's smarter engines look at the title of your site, the meta information that you've included in the Head of your HTML code (we'll get to that), and the actual words that are on your page. If you put "Denver Broncos" in your title and meta info, but your web page is about how to fix a sink, the search engine knows something is wrong. It won't give you a good listing.

All this means that it's easier than ever for busy business folks to put together a web site that search engines will like. Here's what to do:

1. Make your web page (or your entire site) closely focused on a topic that can be summed up in a single keyword or two. My site is about "marketing." The title of the page (the name that appears in the little box at the top of your browser), the meta information, and the words on my page all talk about "marketing." When a search engine indexes my site, the computer has no problem figuring out that my site really is about "marketing"...and there's LOTS of mentions of "marketing" there.

2. Different search engines focus on different aspects of your site, but most place a heavy emphasis on your title-- that line in the box on your browser. Be sure to include your most important key word. Some people like to include it twice if they can use it in a logical sentence. I could use "Nunley's marketing site: free marketing information." Of course, going too far with search engine tactics can make your site read and look funny.

Collection of DrNunley.com Marketing Articles

3. Several search engines put heavy emphasis on your meta information. That's a line in your page's HTML code that gives the engine additional information on the topic of the site and keywords that correspond with what's in the text. It looks like this:

<HEAD>

<TITLE>Dr. Nunley's Marketing Info Supersite!</TITLE>

<META NAME="description" content="Dr. Kevin Nunley's Marketing and Advertising Supersite-dozens of articles on marketing, advertising, and media for small business by one of the Net's top writers.">

<META NAME="keywords" content="marketing,Marketing,marketin,MARKETING,Internet marketing,on-line marketing,advertising, media, ads, copy, copywriting,commercials">

</HEAD>

You can use this same code. Simply remove my title and put in yours. Then replace my keywords with ones that describe your page. Notice that I've used "marketing" several different ways, including one common typo that people often make when typing "marketing." Don't get too carried away with using one keyword. Keep it down to seven times at most (otherwise the search engine will disregard the keyword).

Some search engine experts are now advising NOT to repeat a keyword in any form or fashion. Engines are starting to penalize for that. Many top sites now simply list seven or so keywords and leave it at that.

4. Search engines can't yet read pictures (even the smartest computers still get human faces confused with pictures of pie!), so provide lots of copy that talks about your main theme and keywords. In other words, make your site about what your title and meta info claim it's about.

All this makes it harder for web designers to trick search engines. In a way, that's good for those of us who are too busy doing other things to become experts in search engine registration. There's a simple formula for success: Design a site that is full of good information on a particular topic, and give the site a name that clearly and accurately describes it. That's good marketing, too.

Now I know you're in a hurry, so you'll be pleased to know that 80% of the people using search engines go straight to one of the six biggest:

Alta Vista: http://altavista.com/
Excite Search: http://www.excite.com/
InfoSeek: http://www.infoseek.com/
Lycos: http://www.lycos.com/
WebCrawler: http://www.webcrawler.com/
Yahoo! http://www.yahoo.com/

Here's a tip....while Yahoo is hard to get listed on, they use the same database as http://www.hotbot.com. That's right, get on Hot Bot and you will automatically be on Yahoo.

Right now you can register with the first six with one click at http://www.all4one.com

Go to each engine and look for the link that says "add URL." For Yahoo, you must first go to the listings of sites like yours, and look for the "suggest a site" link on that page.

I also advise registering with AOL Netfind. AOL's 11 million members make it the single largest window to the Internet.

William R. Stanek, author of the book "Increase Your Web Traffic in a Weekend," has provided two handy places to register with many more major link libraries and business directories:

http://www.tvpress.com/promote/yellp.htm
http://www.tvpress.com/promote/guide.htm

Granted, I've tried to explain search engine registration in simple terms. There are many more insights and nuances you can explore (a whole industry has grown up around search engine manipulation). But following these simple guidelines will ensure that your web site is search engine-friendly. You will be much more likely to receive a favorable listing that will dive many more prospects to your web site.

Collection of DrNunley.com Marketing Articles

IF AT FIRST YOU DON'T SUCCEED

Five Ways to Fix Slow Business and Give Yourself a BIG Second Chance.

This would normally be a busy time of year for Margaret, but business is slower than usual. She worries things will get ever slower in the months ahead. Greg came up with a terrific idea for building a second income from the Internet. Months later, his web site and advertising have only brought in a few sales. He is afraid all his time, money, and enthusiasm were wasted.

I hear similar experiences from dozens of people each week. Some are start-ups, others are mature businesses. Most business ideas flop on the first try. The key to success is knowing how to give yourself a BIG second chance. Sometimes you will need to try a third or fourth time before your new product or service brings home the bacon.

Here are five simple ways to give your business new life.

1. Give your business a tighter focus.
Many businesses are too broad, trying to interest too many different kinds of prospects. Being too general will leave you lost in a crowd. America has more stores than at any time in history. Retailers are finding their markets split into tiny fragments as shoppers have a bewildering choice of places to spend their money. Meanwhile, the Internet is exploding. Christmas spending in 1998 is 2.3 Billion dollars, twice what it was in 1997. All that money is divided among some 300 million web sites.

How do you compete when there are so many others? Tighten your business focus to include a narrow, very well defined audience. The man who sells John Denver memorabilia from the 1970s has a very specific, almost peculiar business. But he is selling his product like crazy on the Internet. He is filling a niche that deeply interests a particular group of people.

2. Make your prices more competitive.
For the past few years consumers have told us they want quality and service with price being much less important. The tightening of the economy has changed that. Now consumers are ranking price as one of the most important reasons they buy from one business and not from another.

Think of ways you can tighten your belt or redefine your product or service to offer it at a lower cost. Maybe you can limit your service to fewer, but still important features. Perhaps your prices are already lower than competitors. You just need to emphasize your lower prices more in your marketing. Lower prices are suddenly an important way to get people to buy.

3. Choose the product or service that sells best for you, then expand it.
Go wide and deep. Offer more versions of the same product or service. If the green one is selling well, come out with a red one and a blue one to offer along side your start performer. Look for more related products or services you can offer. I write press releases for people. I have also found those same people

Collection of DrNunley.com Marketing Articles

want me to write articles for them. That is a related service I can offer along with the popular press release service.

4. Sharpen your marketing materials.
With all that competition in the business world, you profit when your marketing and advertising stands out and hits home with consumers. Give all your marketing pieces a headline. Busy prospects need a way to quickly find out "what's in this for me" before they will take the time to read your sales letter, brochure, classified ad, or web site.

Relate the features of your product or service to the benefits the buyer will get. Your "Widget 900" has a clever lever. Tell prospects how that lever will save them time, money, and make their day more enjoyable. It is the benefits that your buyers really care about.

Take a closer look at where you are putting your advertising dollars. It is tempting to place all your cash into big media that reach a lot of people, but are all those people your best prospects?

Marketing is generally more effective when it can be closely targeted to a well defined audience. If your audience is made up of lots of your best customers, you get sales. Consider advertising in trade publications, email newsletters, and neighborhood papers. Postcards are cheap and prospects read them without having to open an envelope.

5. Expand your promotional effort. It takes a LOT of advertising, marketing, and promotion to get into the heads of your busy prospects. People are constantly bombarded with ads and commercials. You need to hit your best prospects over and over again before your message sinks in.

Look for several ways you can CONSISTENTLY market your business. Find affordable methods that reach your best customers and use those methods over and over again. When marketing doesn't work, it is almost always because the business ran out of money and gave up too soon.

Give your business a big second chance. The children's story of the "little engine that could" might as well be a $1,000 business seminar. The best way to clobber competition and build your business into a cash cow is to give your business a tight focus, make prices more competitive, expand what works, improve marketing materials, and promote big and consistently.

Collection of DrNunley.com Marketing Articles

GET LISTED ON SEARCH ENGINES BY GOING THROUGH THE BACK DOOR

Why your site isn't getting listed and what you can do about it NOW. Get all the basics...and some crafty insider strategies, too.

DrNunley.com Search engines can be an incredible form of FREE advertising. Get your site listed high on a major search engine and your hits can zoom to 2,000 visitors a day. Sales will soar overnight!

Unfortunately you, me, and many others have had a hard time getting that elusive search engine listing. You may have tweaked your site for what search engine spiders are looking for, then carefully registered your site with dozens of engines, then waited weeks while nothing happened.

Let me give you some surprising reasons why your listing isn't getting seen. I'll also let you in one some insider ways to easily get in the back door of key search engines.

Uh Oh! We lost your listing.
Last fall Alta Vista, which had just bumped past Yahoo in popularity, suddenly dumped millions of listings from their database. Reports say they accidentally reformatted a few key hard drives. OOPS!

If your hits started drooping, the Alta Vista problem could be the explanation.

Meanwhile, several major search engines where taking your URL submission and would promise to index your site. We now know they were unceremoniously dumping those submissions in the trash.

If you went back and tried to find your listing and it wasn't there, you probably were not listed.

And Yahoo is another story. While they attract a full 55% of search engine users, their directory is notoriously hard to get listed on.

Check your listing.
Take a moment to see if you are currently listed in major search engines. Expert Jerry West says there are really only three that you need to be concerned with.

"Type your web address into Yahoo, Alta Vista, and Excite. Those three account for 88% of all search engine traffic. If your site comes up, you're listed," West says.

I checked to see if my DrNunley.com was listed. Because I registered with all of them last month, I figured I would pass with flying colors. Wrong. Alta Vista had us listed fine, Excite didn't have us listed but offered me wallpaper with my domain printed on it, and Yahoo came up with a site I discontinued two years ago.

In through the back door
You can EASILY get listed on several major search engines by going through the back door. Yahoo and several others get their listings from the Inktomi database. Get listed on that database by submitting your site to Canada.com. They will have you up in three days or less. By default, Yahoo will have you listed, too.

Also hop over to dmoz.org. This is the Open Directory Project begun by several key players from Netscape. It's a snap to get listed and their database is the one Lycos and Hotbot use.

By all means RE-register with Alta Vista-- three or four times. Alta Vista has several databases they rotate (which explains why your site is listed high one minute and not the next). Submit your URL every week for four weeks to cover the entire system.

While you are at Alta Vista, check their short tutorial on how to write Meta Tags for your pages. Find it on the "add url" page Several engines use Meta Tags to list your site.

Take time to write an appealing description for your site to go in your Meta Tags. This is the description that appears next to your link on search engines. Even if you aren't ranked tops, a good description will draw plenty of traffic.

Make sure the words people use to search for you are listed in your page Title and again in your page's copy. If a search word appears in the title but NOT the copy, the engine will ignore it.

Keep up on the latest developments with search engines at SearchEngineWatch.com. By all means, check out Jerry West's excellent reports on search engines at http://WebMarketingNow.com.

Collection of DrNunley.com Marketing Articles

HOW TO BUILD YOUR BUSINESS EMPIRE ON THE INTERNET FOR NEXT TO NOTHING.

The Internet is truly a remarkable opportunity. Anyone, absolutely anyone can harness a major mass media to spread the word about themselves--and do it almost for free. Build your website, promote it like crazy, and you're on your way.

Is creating an Internet empire an easy thing to do? In many ways it is, but you have to have a plan. And, you have to work your plan, consistently, with a determination not to give up until your reach your goals. Internet business is like any other business, success requires hard work and persistence.

TARGET YOUR MARKET.
Increase your odds and shorten the time required by planning carefully right from the beginning. Start by finding a solid product, service, or idea that you can sell to others. Make sure the people who would want what you sell can be found on the Internet. The on-line crowd is still just a small percentage of our total population (although that's a lot of people in total numbers). Netizens tend to be above average in income and education. Until recently, they were overwhelmingly young and male, although the number of women and people of all ages rapidly increased last year. People who buy on the Net like to use credit cards. Can your business sell to these groups of people? Does your product or service appeal to them, or would it sell better to other groups who aren't using the Internet yet?

Once you zero in on a perfect product or service to sell, consider what other products or services could be sold along with it. There's a good reason why you need to sell things that can be grouped together.

PROMOTING AN IDENTITY.
The success of your Internet business will depend almost entirely on how well you can promote it. While traditional businesses throw fortunes at radio, TV, and newspapers; the on-line entrepreneur must do most of the marketing herself. Properly using newsgroups, free classified ads, solicited mailing lists, e-zine ads and press releases is hard enough to do with one busness. You don't want to have to divide your time between two, three, or four different websites. It's much better to have one multi-faceted website to send all your prospects to.

Work toward creating a single, recognizable Internet identity for yourself. Everyone knows Amazon.com. They sell books.

They don't sell financial services, health items, and travel packages--just books. But Amazon.com sells many kinds of books and a number of different related services including an entrepreneural opportunity. It all fits under their hotly promoted identity.

THREE STEPS FOR CREATING YOUR IDENTITY.

Here's how to hone in on your potent Internet identity:

1. **What is the main benefit that your business will offer prospects?** People are only interested in the features your product or service has to offer if they can clearly see how those things benefit their lives. I could have titled this article "An Analysis of Internet Targeting." That sounds kind of dry, and unless you're a marketing student, it probably doesn't mean much. By calling this article "How to Build Your Business Empire on the Internet For Next to Nothing," I communicate some real benefits to reading the article.

   I could go one step further and add to the title "and make a huge income without having to have a boss, a regular job, or leave the comfort of your own home." Now those are benefits that many of us can get enthusiastic about!

2. **Boil your main benefit down to a short phrase you could put on a bumper sticker.** My business's bumper sticker ID is "Marketing help for biz." Most prospects are leading very busy lives. To get their attention, you must tell them how to improve their lives quickly and in a way that's easy to understand.

3. As you promote your business, and multiply the ways to make profits in your Internet empire, **make sure that everything fits with your main bumper-sticker positioning statement.** One guy has a list of 70 money-making opportunities on his website. He has them collected under a concise promise of supplying readers with dozens of ways to make money on their own websites. It's a good way of grouping lots of different business strategies under one unified heading and and identity. It's appealing, has lots of obvious benefits, and is easy to promote with a short bumper-sticker phrase.

You can't start too soon thinking about how you will market your business. When most people get an idea for a new business, they first think about how much money they'll make. The very next thought should be how they will promote the business. As you create and expand your Internet Business Empire, keep a constant eye toward how you will marketing it, efficiently and effectively.

Collection of DrNunley.com Marketing Articles

FIVE WAYS TO GET PEOPLE TO TRUST YOUR ON-LINE MARKETING

One of the Internet's biggest problems is that many people don't quite trust it yet. All new forms of media go through this early in their existence. When telephones first came out, people were scared to death that scoundrels would use them to steal their daughters.

Early radio stations were so alarming to the public that governments around the world practically tripped over themselves hurrying to enact stern protections.

The Internet is having some of the same problems. When the general public is asked why they don't buy more things on-line or from businesses that market on-line, many admit that they're afraid of what lurks behind this incredible new technology.

Of course, much of this is simply fear of the new and it will gradually disappear. In the meantime, here are five simple, but very powerful, things you can do to help prospective customers trust your on-line marketing.

1. **Tell readers about YOU.** Include yourself in everything you do on-line. People want to know WHO is behind the sales letter, the web site, the product line, and the offered service. Don't be modest. Supply prospective customers with lots of details about you, your business, and how your business got started.

   Let readers know why you do what you do. Putting yourself into your marketing gives your on-line advertising a human touch. When readers feel they know you, they begin to trust you.

   Ruthie sells her custom made afghans from her web site. Each page includes her photo in the corner. Her grandmotherly image, smiling at the prospect, helps to put on-line shoppers at ease.

2. **Give full details about your offer.** Don't leave people guessing about what you're selling. Rather than reading through five pages to find out what you're up to, the vast majority of readers will click away if they think you are trying to confuse them or have something to hide.

   Tell people right from the very beginning what you are selling.

3. **Stay away from cliched marketing.** Many people associate copy that starts with "I threw it away" and "Read this twice, then read it again" with get-rich-quick schemes sent as unsolicited e-mail. It's much better to start your on-line marketing with a headline that outlines the most enticing details of your offer. Then quickly fill in the rest of the basics. After that you can include full information for the reader that wants all the facts she can get.

4. **Include a guarantee to reduce the risk of buying.** A 30 day money-back guarantee is required for mail order items. Some on-line businesses have

extended that guarantee to 60 days or even a full year. It's hard not to trust a company that stands behind their products and services for that length of time (and in practice, very few people will ever ask for a return or refund).

5. Supply prospective customers with more traditional ways of contacting you. Give phone numbers, regular mailing addresses, and include the names of principle members of your company. This reassures customers that you are a "real" company and your claims can be trusted. Such disclosure is also being required in state legislation and proposed regulations.

In the end, few things are as valuable to a business as customer trust. That's why famous trademarks and widely known franchises are so important. Their names, products, and services are familiar and trusted by the public.

Work to achieve a personal touch in your on-line marketing. Be clear about what you're offering and provide reliable service. You'll be giving prospective customers plenty of reasons to trust you.

Collection of DrNunley.com Marketing Articles

Four Part Series:
THE INTERNET: WHERE FUTURE BILLIONS WILL BE MADE. PLUG YOURSELF IN!

There is now no doubt. The Internet really IS the biggest gold rush of our lifetime. It is unlikely you or I will get another chance as big as this one to earn huge profits anytime in the next 100 years. The World Wide Web is booming! Internet sales are growing by leaps and bounds, and even a dipping world economy can't make online entrepreneurs slow down or look back.

Ninety-two million people in the US and Canada surf the Net. That's a whopping 16% jump in just nine months. People aren't just looking, they are also BUYING online. The number of online consumers has increased 40% in recent years to 28 million. Perhaps even more important, the number of women buying online went up 80% in the past nine months reaching the 10 million mark.

It is time to get on board! This article tells you how to get online, get an Internet presence that SELLS, and get your piece of the new Internet pie. It is time for you to stake a place on the Net for yourself, your children, and the generations that follow. Someday people will look back and judge us as one of two groups: those who didn't recognize the digital revolution and missed the greatest chance of our age, and those who smartly made a place for themselves in a new business model that will dominate the future.

RUN A PERFECT HOME BUSINESS
You don't have to have a website to run a successful home business on the Internet. E-mail is the Net's most powerful and popular feature.

You don't need an office in a prestigious address. Some of the Net's biggest earners work from a corner in their bedroom, still in their pajamas. Best of all, you don't need a wad of cash to get your home business ideas rolling. Hard work, dedication, vision, and a dream you are willing to see to the end are far more valuable than lavish start-up funds.

There are an endless number of ways to make money on the Internet. You can offer a service delivered conveniently by e-mail. You can take orders for products that are shipped to customers in the US and around the world. It's possible to start your own e-mail newsletter and get paid for running other people's ads. Building a hot website and earning money by letting companies display their banners on your pages are also options. You can also earn a good extra income simply by referring customers to other companies with a link on your website or in your email messages.

NICHE YOUR MARKET AND PICK YOUR PRODUCTS AND SERVICES
The Internet is a big place. Millions of people are shouting about billions of products. Online advertising is low-cost, but it is also a little bit like yelling your message in a crowded football stadium. Try to advertise to everyone, and you will wind up getting the attention of no one. It's important to niche your marketing.

Begin by deciding exactly who will be your best prospects and customers. If you don't already have something to sell, start by looking for a big group that is online and needs help.

After trying lots of things online, I found a huge number of people needed help writing copy. Many people don't like to write. They find they can't get started without web copy, sales letters, and ads. Lots of folks are more than happy to pay to have their copy written for them. They have a problem, and are looking for someone to sell a solution.

Your best customers will be people who have pressing needs or problems you can solve. They hurt. You heal.

Look for places your target audience goes to find products and services. What kinds of websites do they visit? What e-mail newsletters do they subscribe to? What magazines do they read? These are the media outlets you will use to market to your best prospects and customers.

The more you know about your target audience, the better you will be able to reach it, and the bigger your sales will be. You can get information on customers by talking to them and by offering a questionnaire on your website or by e-mail. Give a free gift or special discount to those who complete your questionnaire. Good, precise targeting quadruples the effectiveness of your advertising and gets your cash register ringing.

Make your e-mail or website copy talk directly to your best customers.

Use easy words and short sentences. Customers love to hear the benefits they will get after they buy from you. Talk about how the customers will save more, earn more, save time, or feel better. Provide examples or statistics that prove your claims. Include testimonials from satisfied customers.

Collection of DrNunley.com Marketing Articles

Four Part Series:

PROMOTE ON THE NET:

A big list of ways YOU can get low cost and FREE marketing on-line.

Making money on the Internet requires a great deal of promotion. The cost is so low that you can create a splash on the Web with no more than a contribution from the grocery money. This makes the Net the perfect place to market your home-based business.

Net promotion is fun. Many entrepreneurs love to promote far more than they like the day-to-day tasks of running a home business. Some jump from one new idea to another just so they can devise new promotional campaigns.

The Internet's most successful home workers take a Barnum and Bailey approach to marketing, having a field day with the Net's big selection of free and low-cost promotional tools. One person with online savvy can do the work of 10 people and spend almost nothing in making his or her name a household word.

Marketing expert Jay Conrad Levinson has adapted his Guerilla Marketing approach to the Internet. His creative techniques are a perfect match for the Internet's do-it-yourself environment. Levinson's book "Guerilla Marketing Online" details a variety of ways you can professionally promote your site for next to no money.

Place your ad on the Web's thousands of free classified ad sites. Type "free classifieds" into any search engine for a complete list. There are several free and low-cost programs that help you automate some or all of the task of posting your ads.

Banner ad advertising is one of the top ways corporate web sites promote themselves. There are now many low-cost packages available for home-based businesses. Packages are either priced on a per click-through basis or on a per impression model. Per click-through buys guarantee you will get a set number of people to click on your banner and go to your web site.

Buy e-zine ads. E-zines, or e-mail newsletters, are the Internet's most affordable advertising media options. They pack a lot of power for very little money. Depending on the publication, ads are typically 3 lines to 50 words, costing between $20 and $40. Many reach 40,000 to 300,000 people. Even a small circulation e-zine can pull good results if it is targeted directly at your best customers.

Start your own e-mail newsletter. Nothing creates sales like your own Internet community. A big group of people who know your name, like you, see you as a person with good information, and hear from you often will provide a lasting, steady source of sales and revenue. E-mail newsletters are by far the best way to build your Internet community. They are easy to produce and are almost free to send.

Understand that most readers are busy, so keep your newsletter short. You can design your newsletter in any word processor. Keep lines about 65 characters long with a hard carriage return at the end of each line (hit enter). This will keep your lines from breaking up in your subscriber's e-mail reader. Get subscribers by putting a sign-up form on each page of your website. Also list your newsletter on all your correspondence and printed materials.

The vast majority of e-zines are free to subscribers. Earn revenue by offering to run ads for your subscribers and other businesses. Charging ten dollars per 1,000 subscribers is one way to price your ads. You can also list your newsletter ad offer in a growing number of e-zine directories.

Swap ads with other e-zine publishers. You will get a lot further on the Internet if you are constantly looking for ways to join efforts with other entrepreneurs. Many home-based businesses have built large subscriber bases for their e-zines by trading ads with other newsletters. Simply send a note to other publishers saying you will run their 3-line ad in your newsletter if they will run your ad.

Through all your efforts to build your e-zine, you will be collecting subscribers. This "opt-in" e- mail list is the most powerful marketing asset you have. Never miss an opportunity to ask someone if they want to receive your information periodically. Add them to your growing list. Send out announcements of new products, lower prices, or new affiliations with other firms.

Your e-mail offers should contain links to pages on your web site that will give serious prospects all the additional information they want. Most e-mail programs now turn any address with http:

in front of it into a live, clickable link.

Use opt-in e-mail. This involves using commercial lists of people who have ASKED to receive messages about a particular topic. Suppliers of opt-in e-mail sell lists of people wanting e-mail on everything from business opportunities, to cars, to tax tips. You can send an e-mail message to several hundred people for as low as $40. Opt-in e-mail can often generate a success rate up to 17%, which is very high for direct marketing.

Don't spam. Sending e-mail to people you don't know is called spamming. It is very unpopular with Internet users with some aspects being illegal in some states. Spamming is very bad for business and can create lots of problems including the cancellation of your Internet connection. Get visibility in Internet Malls. These are large groupings of websites covering particular topics or product categories. There are Internet malls for business opportunities, for crafters, and for businesses located in specific geographical areas. People interested in your kind of products and services can find you in these handy one-stop-shop locations.

A good mall can help you get your Internet business started with expert advice and easy-to-use credit card transactions for your customers. You don't have to spend big money, learn complicated programming, or hire more people to do handle the details.
An Internet mall should receive plenty of promotion from its owners. Like anything else on the Internet, a mall needs to be aggressively marketed to get visitors.

Collection of DrNunley.com Marketing Articles

Four Part Series:

GET A WEB SITE THAT SELLS!

Ways to make sure your site turns visitors into customers

On the Internet, a good website can make you as big as any corporation. You can get a good designer to come up with a logo, some buttons, and a background that looks right for your business. Then build your own web pages around the elements the designer has given you. This is an excellent way to get an eye-popping web site for as little as $100.

Don't get too carried away with bright colors, patterned backgrounds, or big graphics that take lots of time to load. Recent surveys show that over 78% of your customers will be using slower dial-up modems past 2002. Slow-loading pages discourage visitors who often click away before they have even had chances to consider your offer. If your site is getting lots of hits, but not making any sales, slow-loading pages could be causing the problem.

Basic colors and a white background do best with the many different ways your page will look on different monitors. Use headlines and subheadings to give customers a quick idea of what your page has to offer. Someone in a hurry should be able to read your headlines and subheadings to quickly understand what products, services, and benefits they will get. Put your most important phrases in bold letters, too. Stay away from all capitals in your writing and use common fonts that are easy to read.

Above all, include testimonials from your customers and suppliers. Nothing builds trust with customers and prospects like good words about you from people who know you and have done business with you.

Your site must be easy to navigate with links to important pages included on every page. The look and feel of your site should be consistent. You can do this by using the same logo, background, type style, and navigation buttons on every page.

A website is the perfect place to display your entire product catalog without having to bear the high cost of printing and mailing. You can build your own professional catalog fast with a few mouse clicks by using the online catalog tool at workshopinc.com/eshowcase.htm and the handy programs at iomanager.com.

Search engines are the number one way of getting visitors to your site. You will want pages that are easily registered by the top search engines like Excite and Yahoo. Search engines are a lot smarter than they used to be. Most first look at the title of your page (those words that appear in the little box on your browser), then at the page's Meta Tags, and finally at the copy on your page. If the same keywords appear in all three places, your site gets a high listing. This means that a customer who searches for you using one of your site's keywords will find you linked in the first 10 to 15 sites the search engine presents.

To see what Meta tags look like, go to any popular site, point your cursor at the page, and click the right mouse button. Choose "view source." The Meta tag looks like this:

<meta name="description"

content="a few words that best describe your site, separated, by, commas">

<meta name="keywords"

content="the same keywords listed like this:internet marketing, small business, web site promotion, selling, ezines, home based business, marketing, advertising, email marketing,">

Make sure you update your important info often. Search engines, just like customers, check back often to see how your site is progressing. Lots of helpful articles, tips, and a question and answer page will give customers all the information they need. Also include links to other recommended sites like yours. Try to get those sites to add your link as well. Customers will see you as an authority in your field and will appreciate your ability to help them in a variety of ways.

Your own domain name will make your business look bigger and your Internet effort more serious. Yourname.com look a lot more impressive than freewebsite.com/10101/yourname. Don't forget to include all your contact information where customers can easily find it. Include phone, fax, and your regular mailing address. Your street address adds credibility to your website and offers.

Invite visitors to email you with questions and for more information.

People on the Web like to make purchases with speed and convenience.

Learn all the details on how you can get your own credit card forms and automate ordering with CGI scripts at HomeBusinessMag.com. Assure customers that their credit card info is secure, safe, and private. Include your fax number, toll free phone number, and mailing address for those who prefer to order via those methods.

After a customer buys, follow-up with an e-mail confirming the order. Tell the customer the price and when the product will be sent. Many corporations are finding this kind of follow-up is the number one reason customers come back to buy again.

One of the best things about a website is that it can make the Internet your extended staff. This is an invaluable boost if you are a home business of one. Your website is there to answer questions at any time of day or night. Your web pages can fill in the additional info that your ad couldn't cover, your sales letter didn't have room for, or you forgot to mention in your telephone conversation. Many top network marketers use websites to train new members and distribute announcements to their downline. They report it works must faster, better, and cheaper than sending out letters or spending hours on long distance.

Include your website and e-mail address on all of your printed and off-line marketing materials. The corporate world spends most of their Internet ad budgets promoting their sites on TV and radio, and in newspapers and magazines. It works. Notice how the company, United Parcel Service (U.P.S.) now features the company website in the same big letters it uses to list its 1-800 number on all of the delivery trucks.

Put your Internet information on your business cards, invoices, letterhead, and voice mail. Look into advertising your site in affordable and targeted magazine ads, on radio, and in cable TV commercials.

Get FREE big media promotion. Almost all e-zine, TV, radio, newspaper, and magazine editors accept press releases. Now that most media sources are online, e-mail is the preferred way for them to receive your releases. Your release should be one page long and offer valuable information of interest to the publications' readers. Editors won't print a blatant ad that is not accompanied by newsworthy information.

Collection of DrNunley.com Marketing Articles

Four Part Series:

HERE COMES BUSINESS... CAN YOU HANDLE IT?

Kevin gives you efficient ways to manage lots of customer questions, orders, and e-mail.
Plus a list of places to get FREE promotional tools!

Once your on-line business starts getting attention and pulling in sales, your work load will increase. Customers will ask questions, orders will fly in, and e-mail will invariably start to stack up. Much of running a successful Internet business has to do with knowing how to use e-mail well.

Use a signature (sig) file on all of your e-mail messages. Thirty million people use e-mail every day, so it makes sense to pack your messages with a special dose of promotion. A sig file is a four to six line message at the end of your e-mail. It contains your business name, a brief line about what you do and what benefit you bring to customers, your website address, and your contact info.

A sig file can be a good and acceptable source of advertising when posting to otherwise non-commercial UseNet Newsgroups. Find a newsgroup that discusses your area of expertise. Post a message offering useful information or answering a question. Your sig file can direct people back to you and your website for additional help.

Handling the rush of response. Now that you have done all this promotion, don't be surprised if you wake up to find your e-mail box brimming with questions and orders from dozens or hundreds of customers. You will need an informed and organized way of smartly working leads and sales.

Follow-up is key to selling. Some customers need seven to 10 contacts with you before they buy. This can increase sales 50%. If you are busy answering e-mail from a flood of new customers, it can be hard to find the time or remember to keep up with those you have already made contact with. Fortunately, e-mail offers a nearly free way to repeatedly follow-up with interested prospects.

Let "smart" autoresponders do the work for you. These simple online gizmos send your sales information to anyone who sends an email to your autoresponder address. The autoresponder then sends your customer additional messages and reminders at intervals which you determine.

A prospect could get your free report on day one, your sales letter on day two, and your special price list on day three. A week later your autoresponder could send them a reminder with another message set to go out next month. All this is done automatically without you ever having to lift a finger.

E-mail programs like Eudora and Pegasus let you quickly route messages into files on your computer. I keep three files titled "Waiting for Order," "Fill Order," and "Bill 'Em." You can also set up templates with answers to questions you get all the time.

Collection of DrNunley.com Marketing Articles

Instead of typing the same answer to a question over and over, you can click a button and the answer appears. This saves you many hours otherwise spent doing e-mail chores.

Market, test, market, test. Few if any ideas work perfectly the first time. Look at any failures as learning experiences and as steps you take toward success. Test your products, services, and marketing. Analyze the results, make adjustments, and try again. You can track the effectiveness of advertising by directing respondents to a special email address or web page. For example, have them reply to info12@yoursite.com or www.yoursite.com/info12 ("info12" could be your code for the ad you placed in a specific e-zine).

STAKE YOUR ONLINE CLAIM!
The Internet is so hot and so unstoppable that leaders in Washington, D.C. are actually having to take steps to slow down the economy. The Net is not only changing the way we do business. It is also changing the way people live their lives all over the world.

Most importantly, the Internet levels the playing field. You no longer have to get a major bank loan or sell millions in stock to build your home-based business empire. Anyone working from home can have a part-time or full-time business earning a steady income on the Internet.

Even if your home-based business is already working well on the Internet, look for ways you can extend your online activities to this global networked community. In the near future, all businesses could be online. Make your move NOW to stake your claim on the bustling Internet frontier.

IT'S FREE!
You can get just about anything you need to run your home-based Internet business without paying a cent. Most of the promotional and administration tools mentioned in this article are offered online at no charge.

Get FREE website space at tripod.com, homestead.com, and xoom.com. These services also include easy programs to help you build your website, even if you have no experience.

SendFree.com offers up to 20 FREE autoresponders and runs your ad at no charge on their big autoresponder network.

Get FREE "smart" autoresponders that automatically send up to 10 return messages at: fastfacts.net, zinfo.net, and smartbotpro.net.

Register your website with over 400 search engines FREE at submit4free.com.

Gary Christensen's big FREE list of e-zine editors looking for articles: site-city.com/members/e-zine-master.

Use Kate Schultz's FREE e-zine builder to quickly turn out a professional e-mail newsletter at e-zinez.com. It even lets you include your ads and tips.

Collection of DrNunley.com Marketing Articles

Top sites for FREE promotional and e-commerce ideas: BizWeb2000.com, WilsonWeb.com, and DrNunley.com.

Send your classified ad to over 2,000 free sites with one click at adlandpro.com.

Chapter 4: Ways to Measure

Counting "Hits" Not Best Measure Of Web Success

Title: Counting "Hits" Not Best Measure Of Web Success
Author: Steven Bonisteel
Abstract: Eric Schmitt, an analyst with Forrester's Commerce Site Research team says using count of page views to determine if a business is performing well on the Web "is like evaluating a musical performance by it's volume.."

Copyright: © 1999 EST, Individual.com, Inc.TM

Charts and graphs summing up "hits" and page views falls short of what it takes to measure success on the Web, according to a report released by Forrester Research Inc.

Eric Schmitt, an analyst with Forrester's Commerce Site Research team says using counts of page views to determine if a business is performing well on the Web "is like evaluating a musical performance by its volume."

In an interview with Newsbytes today, Schmitt said businesses need to know who the people who visit their Web sites are, why they're there, and how those visits fit in to their relationships with those customers both online and off-line.

"I can look at log (reports) and know how many hits and page views I had last month, but did my call volume drop in my call center as a results?" Schmitt asked. "Were the people who came to my site existing customers, or are they new customers? What pathways are they cutting through the site? What works in terms of site design and layout?

Schmitt said coming up with that kind of information is "really a black art right now. And if people are even asking those questions, they're not getting good answers. We're still in the world of pie charts and bar charts that show hits and page vies, and what times of day are busiest."

For its report, "Measuring Web Success," Forrester surveyed 50 large firms to determine how they measure the success of their Web initiatives. Hits and page views represent the most popular means of judging success, Forrester reported, saying that 75 percent of them use low-cost measurement tool packages, yet expressed dissatisfaction with the performance, storage, and trustworthiness of these tools.

The result, Schmitt said, is that companies will attempt to build e-commerce or customer-support sites using

Counting "Hits" Not Best Measure Of Web Success

"personalization" gadgets without having the customer data to back that up.

"People get sold on all the wonderful things a good personalization system can do," he said, "but they don't recognize that underneath every good personalization system is a solid data warehouse. But data warehouses aren't sexy, and people don't want to talk about data warehousing tools, and shop for them and consider the labor that goes into them."

In the end, he said, businesses can end up being "overly aggressive" in the resulting personalization efforts. "They know a little bit about you and then make assumptions that are not correct - and actually end up doing more harm than good.

"If the data is not there, you end up with an engine without gas," he said.

There are a number of reasons for the disconnect between customer information systems and Web sites, Schmitt said.

"For existing companies - not the dot-com startups - the issue is that they have a lot of legacy software systems that were not designed to face an interactive customer," he said. "So, you have a lot of useful information about folks but it's trapped in different data silos. When you go to a company's site they may, for example, try to upsell you on some item not knowing that you already have a trouble ticket in on (that product).

"The effort of synchronizing back-end data sources and putting a customer- friendly face on it has impeded a lot of traditional companies," Schmitt said. "If I am putting in a piece of call-center software today, I'm certainly thinking about what kind of Web functionality it has. But if I put it in three years ago and I socked a half a million dollars, or 5 million, into it, am I just going to rip that out? Most companies can't afford to do that, so they're hamstrung in a way.

"Then, for a lot of the dot-coms, they just don't know that much about people. They're still in customer-acquisition mode. Customer profitability, active customers...those terms don't really apply to them yet because they're still trying to get people to show up in the first place."

Schmitt said he sees the situation changing in the coming year. "It's fair to say we're going to enter an era of accountability," he said.

"Right now we're in 'get-it-up-there' mode. There's this rush to get something out there, but probably this year we're going to see the hard questions being asked, like: 'I just made a $20 million Web investment, now what am I getting out of it?,' 'How much did I cannibalize my (off-line) sales?' or 'How much am I saving in terms of customer-support costs?'"

To support business goals, firms need a robust Web intelligence model capable of accommodating new strategies and technologies, Forrester said. Backstopping all the information gathering will be data warehouse technology that was emerging earlier in this decade before being pushed

Counting "Hits" Not Best Measure Of Web Success

out of the technology headlines by excitement over the Web.

"Data warehousing projects are not new," Schmitt said. "It's sort of a case where the guys who never got any glory in the client-server world (now find) that we need these very same skills on the Web."

Forrester Research can be found on the Web at: http://www.forrester.com

The Dirty Truth About Click Throughs

Title: *The Dirty Truth About Click Throughs*
Author: *eMarketer "the authority on business online"*
Abstract: *Should advertisers pay for the number of eyeballs that visit a page and "see" their banners or the number of fingers that click a mouse on a banner and go directly to an advertiser's site? In the meantime click-throughs rates are falling because:*
1. The thrill is gone,
2. More directed traffic,
3. Banneritis.
Banners have to work harder.

Copyright: © 2000 e-land, inc.

There is debate in the online industry concerning impressions vs. click-throughs. Should advertisers pay for the number of eyeballs that visit a page and "see" their banners or the number of fingers that click a mouse on a banner and go directly to an advertiser's sites?

It is the new media version of an age-old advertising argument: How do you verify the success of advertising (in any medium) -- by its ability to build the image of the brand in the consumer's mind, or its efficacy in driving customer purchase? Or does advertising deliver some mystical combination of both? Beyond that, how do you measure the "worth" of a brand or the impact of advertising on sales?

No one knows. After centuries of adverts, billions of advertising impressions and millions of dollars spent on research, nobody can tell you exactly how or why advertising works. In fact, there seems to be only one point of agreement: Advertising works...somehow.

Without advertising -- whether ads or promotion or awareness or PR or attention-getting packaging -- no product sells. People have to know a product exists before they can buy it, and they have to understand what it does before they will buy it. To have any chance of selling anything, you have to reach your potential customers, get their attention and communicate to them that your product exists and that it has value -- usually more value than other products.

Sounds simple. But in reality it is difficult to do. There are just too many competing products (clutter) and each consumer is busy and pressed for time. Most no longer read Bibles, books or the manual for how to maintain their $85,000 Mercedes, so chances are slim they will stop to read an ad.

Little wonder advertisers are scoffing at web page "impressions," in reality just hits, which could include the same people moving back -- click-click -- across the same page -- or multiple visuals loading.

The Dirty Truth About Click Throughs

That's like a magazine charging for the number of times an ad page gets turned as readers flip back and forth through a magazine.

Here's the kicker:

Click-Through Rates Are Falling.

Over the last year and a half, as online advertising has matured, the accepted "norm" for click-through rates has been 2%. That means for every 100 impressions -- hopefully, visitors -- two would click a page's banner. Above 2% you had a very good execution or offer, below and it was time to reconsider your creative or media buy.

But now that is changing.

Click-through rates are dropping for a number of reasons:

1. The Thrill Is Gone
Not too long ago banners were new, and they weren't up everywhere. It was fun -- a different, unique experience -- for surfers to click on a banner and be whisked to another site. Banners were another way of experiencing the hyperlink-magic-carpet ride of the internet. But -- Hey! -- after going to a few dumb, off-track sites ten or hundred times, the experience loses its charm.

2. More Directed Traffic
The audience characteristics of surfers have changed. Two years ago a majority of the people online were netheads, computer geeks and college kids, looking for a good time as much as anything. Now corporations have roared online. People come online looking for a certain site or a particular bit of information -- they usually don't have time to just surf around willy-nilly.

3. Banneritis
Everyone who has been online for any length of time has clicked on a banner and suddenly ended up somewhere they didn't want to be. An insane message, a slow downloading site, Java code that crashed the browser, all kinds of bad or irritating things can happen when someone clicks a banner out of curiosity. As a result, the audience is getting more discriminating. They don't click banners unless the offer of message is extremely compelling for them.

Banners have to work harder to get less click-throughs.
This is not surprising. A "good" response rate on a mass mailing -- which is certainly no longer a novelty to anyone -- is .05%. That's one fourth of the accepted banner rate, and probably close to what the normal click-through rate will soon be. And that's using sharp, well-defined targeting combined with attention-getting compelling offers, two things most banners don't currently offer.

On-Line Advertising Campaign Measurement

Title: *On-Line Advertising Campaign Measurement: How Cached Impressions and Varying Ad Serving Technologies Affect Reporting and Performance*
Author: *Nicole Goldstein*
Abstract: *Certain counting methodologies – specifically static ads, ad insertion and dynamic ads – ignore cached impressions. The end result is that publishers end up over-delivering impressions in order to make-up for those that are never counted. These impressions never show up on the buyer reports and so, in effect, buyers get more impressions than what they paid for and they receive performance results that are skewed and misleading. Cache busting does help but does not solve the entire problem because impressions viewed when a user hits the "back" or "forward" button on their browser are not always counted, even when cache-busting code is put in place.*

Copyright: © 1998 Adauction.com "Opportunity CLICKS"

Overview:

In general, performance of an on-line advertising campaign is measured by its click-through rate – that is, number of clicks on a banner divided by the number of impressions shown. In an ideal world then, the variables that affect click-through rate would be intangibles like banner creative and audience targeting. Unfortunately, the current state of the World Wide Web is such that basic measurement of impressions has not yet been standardized. The end result is that when all other variables are equal (same banner, same targeting criteria), the performance of an ad campaign can vary significantly from website to website, ad server to ad server. This is because each ad server counts impressions differently.

The confusion is caused by the fact that ad impressions can be served from a computer's memory (cache) as well as from an ad server. Those delivered from a server will be counted; those delivered from cache may or may not be counted.

The purpose of this document is to shed light on the way ad impressions are cached and then to detail how different methodologies are employed to count ad impressions, cached and non-cached. This information is critical in that it will allow us to fully understand why the performance of a single ad campaign can vary so dramatically from one website to another, from one ad server to another.

Caching:

The whole idea behind caching is that it speeds up the internet. Once a user views a web page, the page's images are stored in cache (memory) - in the browser and/or on a proxy server. The next time the user visits the same page, the browser (or proxy server) will recognize the image and will pull it from cache instead of making another call to the server. This minimizes download time and eases traffic on the internet.

This is great when you're a user trying to get information quickly and efficiently.

On-Line Advertising Campaign Measurement

However when you're a website trying to maximize ad revenue or an advertiser trying to obtain true performance statistics, caching skews results. Here's why:

Ad banners served from an ad server will be counted. Those served from cache, by definition, bypass the server and are therefore never counted. From a purely mathematical standpoint, the denominator of the click-through rate equation is lower than it probably should be resulting in a higher click-through rate. On a less theoretical level, more impressions are served then are counted. Because a user can click on cached and uncached impressions and all clicks count, the click-through rate ends up being higher.

On the following page is a diagram of how a page can be cached in the browser and on a proxy server.
The following scenarios highlight how caching effects impression counting when a site is serving directly using one of the four main ad serving technologies. The four technologies are: Static Ads, Ad Insertions (Net Gravity & Real Media use this type), Dynamic, and Cache-Measuring (AdForce and DART use this type of methodology). A fifth scenario, one of the more common encountered by us at Adauction.com, is also detailed.

These scenarios are based on specific user behavior:
1. Two users visit a website's homepage
2. They both click to a new page
3. They both hit the "Back" button on their browser
4. They both hit the "Forward" button on their browser

Assumptions: Both users are using either Netscape or Internet Explorer.

On-Line Advertising Campaign Measurement

Figure 1

Browser Caching and Proxy Server Caching: A couple of general scenarios

First Time Viewing a Web Page

1. User 1 enters URL into Browser. browser checks to see if page is cached.
2. Since this is the first time in, page is not in browser cache.
3. Browser routes request to the proxy server
4. Proxy server checks its cache
5. It's not found in proxy server
6. Proxy server routes request to the website's webserver
7. Webserver sends the page HTML back to the proxy server
8. Proxy server stores page in its cache
9. Proxy server sends page HTML to browser
10. Browser renders page and stores page in its cache

An Example of Browser Caching

1. User 1 hits the "Back Button"
2. Browser checks its cache to see if it recognizes the page. ot does and so it pulls page from browser cache

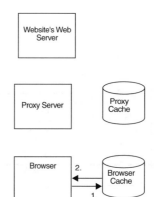

Proxy Server Caching

1. User 2 has never been to the website before. Enters the URL into their browser. Browser checks its cache.
2. Since user 2 has never been to the page before, page is not in cache.
3. Browser routes request to teh proxy server.
4. Proxy server checks its cache.
5. Since the page is in proxy server's cache (user 1's request left it there), page is pulled from PS cache.
6. Proxy server sends page back to the browser
7. Browser renders the page and stores the page in its cache.

The Net-Net of all of this: One page has been counted by the webserver; Three pages have been seen!!!

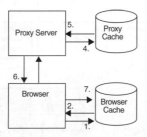

The Ultimate Internet Advertising Guide

Static Ads

Static Ads are ads that are basically hard-coded on a web page. The page has an image file and a click-through URL that changes only when the website takes some action (either modifies the banner HTML or runs a program that replaces the current banner with a different banner). In other words ads on a particular page will not rotate because the page has been hard-coded to display one particular banner. With static ads, ads that reach the web server are counted as "hits". See CASIE "How interactive ads are delivered and the measurement implications" document - http://www.casie.com/PUBS/index.html – for more information.

Table 1

User Goes To:	User	No Cache- Busting			With Page Expiration		
		No. Impressions Counted	Impressions Seen	Impressions pulled from:	No. Impressions Counted	Impressions Seen	Impressions pulled from:
Homepage	User 1	1	Banner A	Web Server	1	Banner A	Web Server
	User 2	0	Banner A	Proxy Cache	0	Banner A	Proxy Cache
Link to Next Page	User 1	1	Banner B	Web Server	1	Banner B	Web Server
	User 2	0	Banner B	Proxy Cache	0	Banner B	Proxy Cache
Back	User 1	0	Banner A	Browser Cache	0	Banner A	Browser Cache
	User 2	0	Banner A	Browser Cache	0	Banner A	Browser Cache
Forward	User 1	0	Banner B	Browser Cache	0	Banner B	Browser Cache
	User 2	0	Banner B	Browser Cache	0	Banner B	Browser Cache
Totals:							
User 1 (browser Cache)		2	4		2	4	
Both Users (Browser & Proxy Cache)		2	8		2	8	
Why this happens:		Because cached impressions are delivered from a computer's memory, the server doesn't know about them. They therefore can not be counted. A site has to actively implement cache-counting/cache-busting techniques in order to count every impression.			A site can not append a random number on a static ad to make the ad unique - if it could, the ad would not be considered static. Therefore, there is no such thing as cache-busting using a random number appended to the banner. Even when a site uses page-expiration, it will have no impact because the banner image will still be loaded from the cache (even though the page isn't). Page expiration with a static ad is valuable if there is an ad on every page and the site can equate page views to impressions. If this is the case, then 8 page views will equate to 8 banners - all impressions will be counted. The scenario above assumes that this is not the case.		
Conclusion:		Results in buyer's favor - buyers get more impressions than what they've paid for. Website has effectively diluted their CPM.			The only way static ads can be cache-busted is to equate an ad to every page view and then to add a 0 second expiration tag to every page. Otherwise, impression counting will result in the buyer's favor.		

On-Line Advertising Campaign Measurement

Ad Insertion

With the Ad Insertion methodology, an ad decision maker decides which banner to display right after the web server has been called. The ad's HTML is inserted directly onto the page as the page is being rendered. The impression is counted at the time the decision maker decides which banner to display. See CASIE "How interactive ads are delivered and the measurement implications" document - http://www.casie.com/PUBS/index.html – for more information.

Table 2

		No Cache- Busting			With Cache- Busting - Random # appended to banner		
User Goes To:	User	No. Impressions Counted	Impressions Seen	Impressions pulled from:	No. Impressions Counted	Impressions Seen	Impressions pulled from:
Homepage	User 1	1	Banner A	Web Server	1	Banner A	Web Server
	User 2	0	Banner A	Proxy Cache	1	Banner A	Web Server
Link to Next Page	User 1	1	Banner B	Web Server	1	Banner B	Web Server
	User 2	0	Banner B	Proxy Cache	1	Banner B	Web Server
Back	User 1	0	Banner A	Browser Cache	0	Banner A	Browser Cache
	User 2	0	Banner A	Browser Cache	0	Banner A	Browser Cache
Forward	User 1	0	Banner B	Browser Cache	0	Banner B	Browser Cache
	User 2	0	Banner B	Browser Cache	0	Banner B	Browser Cache
Totals:							
User 1 (browser Cache)		2	4		2	4	
Both Users (Browser & Proxy Cache		2	8		4	8	
Why this happens:		Because cached impressions are delivered from a computer's memory, the server doesn't know about them. They therefore can not be counted. A site has to actively implement cache-counting/cache-busting techniques in order to count every impression.			The random number associated with each banner forces the banner to appear as a unique banner. So, the first user will see a banner which will be counted. This banner will be stored on the proxy server. When the second user comes in for the first time, the ad banner on their page will have a different unique number - hence, that banner will be counted. This also occurs whe each user views different pages within a website. In effect, the random number appended to the ad ensures that a different banner is displayed as users move from one page to another within a website. However, as soon as a user presses the "Back" or "Forward" button on their browser, the page will be pulled from the browser cache - the inserted ad will already be embedded on the page and the random number will be one that the browser has already seen. So, the ad decision-maker won't be called again.		
Conclusion:		Results in buyer's favor - buyers get more impressions than what they've paid for. Website has effectively diluted their CPM.			The random number defeats some caching but not all - buyers still get the benefit.		

Dynamic Ads

Sites that serve Dynamic Ads place a generic HTML tag onto their pages. This tag makes a call to an Ad Server which in turn decides which banner to display. The impression is counted when the ad is delivered to the browser from the ad server. See CASIE "How interactive ads are delivered and the measurement implications" document – http://www.casie.com/PUBS/index.html – for more information.

Table 3

		No Cache- Busting			With Cache- Busting Random # in tag		
User Goes To:	User	No. Impressions Counted	Impressions Seen	Impressions pulled from:	No. Impressions Counted	Impressions Seen	Impressions pulled from:
Homepage	User 1	1	Banner A	Ad Server	1	Banner A	Ad Server
	User 2	0	Banner A	Proxy Cache	1	Banner A	Ad Server
Link to Next Page	User 1	1	Banner B	Ad Server	1	Banner B	Ad Server
	User 2	0	Banner B	Proxy Cache	1	Banner B	Ad Server
Back	User 1	0	Banner A	Browser Cache	0	Banner A	Browser Cache
	User 2	0	Banner A	Browser Cache	0	Banner A	Browser Cache
Forward	User 1	0	Banner B	Browser Cache	0	Banner B	Browser Cache
	User 2	0	Banner B	Browser Cache	0	Banner B	Browser Cache
Totals:							
User 1 (Browser Cache)		2	4		2	4	
Both Users (Browser & Proxy Cache)		2	8		4	8	
Why this happens:		With Dynamic Ads, everytime a page with the generic tag is rendered from the webserver, the impression will be counted. If the page is cached and there is no cache-busting on the tag, then the tag will be cached as well. If the page is expired but there is no cach-busting on the tag, the page will be served from the server (and therefore counted) but the tag will still be cached and not counted.			Appending a random number to a dynamic ad helps a little bit. As a user moves from page to page, the random number ensures that the ad server is called each time. However, once the user hits the "Back" or "Forward" button, if the page has been cached it will not regenerate the random number. So, the browser will pull the banner out of cache and the impression will not be counted. The way to truly defeat cache in this instance is to expire the page every 0 seconds (this ensures that the page will always be rebuilt) and add a random number to the tag (this ensures that every page rebuild generates a new random number which forces the ad server to be called).		
Conclusion:		Results in buyer's favor - buyers get more impressions than what they've paid for. Website has effectively diluted their CPM.			The random number defeats some caching but not all. Results are still skewed in favor of the buyer - more impressions are seen than are counted.		

Cache-Measured Ads

Like Dynamic Ads, sites that use the cache-measuring technique place a generic HTML tag on their web pages. This tag still calls an ad server but instead of returning a banner back to the browser, the ad server decides which ad to display and then returns a set of instructions to the browser. These instructions tell the browser where the image is located. The browser then redirects to the image location, pulls the banner, and displays it. The impression is counted at the point the ad server is called. The image can be served either from the server or from cache. See CASIE "How interactive ads are delivered and the measurement implications" document - http://www.casie.com/PUBS/index.html – for more information.

Table 4

User Goes To:	User	No Cache-Busting			With Cache-Busting Random # in tag		
		No. Impressions Counted	Impressions Seen	Impressions pulled from:	No. Impressions Counted	Impressions Seen	Impressions pulled from:
Homepage	User 1	1	Banner A	Ad Server	1	Banner A	Ad Server
	User 2	1	Banner A or B	Ad Server or Proxy Cache	1	Banner B	Ad Server or Proxy Cache
Link to Next Page	User 1	1	Banner C	Ad Server	1	Banner C	Ad Server
	User 2	1	A, B, C or D	Ad Server or Proxy Cache	1	Banner D	Ad Server or Porxy Cache
Back	User 1	1	Banner A, B, C, D or E	Ad Server or Browser Cache	1	Banner A, B, C, D or E	Ad Server or Browser Cache
	User 2	1	Banner A, B, C, D or E	Ad Server or Browser Cache	1	Banner A, B, C, D, or E	Ad Server or Browser Cache
Forward	User 1	1	Banner A, B, C, D or E	Ad Server or Browser Cache	1	Banner A, B, C, D or E	Ad Server or Browser Cache
	User 2	1	Banner A, B, C, D or E	Ad Server or Browser Cache	1	Banner A, B, C, D or E	Ad Server or Browser Cache
Totals:							
User 1 (Browser Cache)		4	4		4	4	
Both Users (Browser & Proxy Cache)		8	8		8	8	
Why this happens:		Whenever the generic tag embedded on a web-page is rendered, it will call the ad server. So, as users navigate through a site that uses cache-measuring, each banner that they see will be counted. This is true even when the user hits "Back" or "Forward" on their browser. In addition, because the ad server is being called, it will serve the banners evenly and appropriately.			You don't need to cache-bust the tag if you use an ad server that "measures cache". The results are the same whether you bust cache or not. This is because the location of the previous image shown is always thrown away and never stored in cache. This forces a call to the ad server.		
Conclusion:		All impressions that are delivered are counted. This is fair to both the publisher and the buyer.			All impressions that are delivered are counted. This is fair to both the publisher and the buyer.		

On-Line Advertising Campaign Measurement

Scenario 5 - The Most Common Adauction.com Scenario

Adauction.com uses third-party ad-servers (AdForce and DART) to deliver what we sell at our auction. Most of the sites for whom we sell inventory use off-the-shelf, in-house ad-servers like Net Gravity and Accipiter. In order to serve our ads, we send a tag (either an AdForce or a DART tag) to the website. They put our tag into their ad server as a redirect. So, a user will enter the URL of the site into their browser. As the page is rendered, the site's ad server (say Net Gravity) is called. Net Gravity pulls our campaign. Our campaign in Net Gravity is actually our AdForce tag. So the AdForce tag calls the AdForce ad server, which in turn selects your banner. The banner is then displayed in the user's browser. The following scenario shows how caching effects a redirect when the site is using NetGravity to serve the AdForce tag. Assume that the site is cache-busting their call to Net Gravity. Diagram B displays the concept of a redirect in more detail.

Figure 2

Diagram B:

A Redirect*

1. User enters the URL of a website into browser. Browser contacts website

2. As web page is rendered, website call it's ad server (Net Gravity). An impression is counted.

3. Net Gravity determines that the next ad campaign to serve is an AdForce campaign.

4. The contents of the page and the AdForce tag are passed back to the browser.

5. The browser contacts the AdForce ad server.

6. AdForce selects a campaign, counts an impression and passes a banner and a click-thru URL back to the browser.

* This scenario assumes that nothing is cached

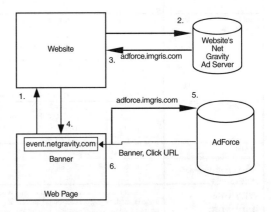

On-Line Advertising Campaign Measurement

Table 5

User Goes To:	User	No Cache- Busting in AdForce tag				With Cache- Busting in AdForce tag			
		No. Imps - Net Gravity	No. Imps AdForce	Imps Seen	Impressions pulled from:	No. Imps - Net Gravity	No. Imps AdForce	Imps Seen	Impressions pulled from:
Homepage	User 1	1	1	Banner A	Ad Server	1	1	Banner A	Ad Server
	User 2	1	0	Banner A	Proxy Cache	1	1	Banner B	Ad Server
Link to Next Page	User 1	1	1	Banner B	Ad Server	1	1	Banner C	Ad Server
	User 2	1	1	Banner B	Proxy Cache	1	1	Banner D	Ad Server
Back	User 1	0	0	Banner A	Browser Cache	0	0	Banner A	Browser Cache
	User 2	0	0	Banner A	Browser Cache	0	0	Banner B	Browser Cache
Forward	User 1	0	0	Banner B	Browser Cache	0	0	Banner C	Browser Cache
	User 2	0	0	Banner B	Browser Cache	0	0	Banner D	Browser Cache
Totals:									
User 1 (browser Cache)		2	2	8		2	2	8	
Both Users (Browser & Proxy Cache		4	2	8		4	4	8	
Why this happens:		OK. The site has implemented the third-party tag as a redirect but the tag is not cache-busted. So, the first time in, Net Gravity is called and an impression is counted on its end. It determines that it must serve the AdForce tag. The AdForce tag calls the AdForce ad server, an impression is counted on its end, and a banner is served. Both Net Gravity and AdForce have each counted one impression. When the second user views the site's homepage, Net Gravity will be called (and an impression will be counted); Once again, Net Gravity determines that the banner to display should come from AdForce. However, since User 1 saw the AdForce tag, it is now cached on the proxy server. The banner will be pulled from proxy server cache and AdForce will not count the impression. This pattern continues - the first user to view a page will see a banner that is counted by both ad servers. The second user to view the page will see the same banner because it will be pulled from proxy cache. This banner will be counted by the site's ad server, but not by the third party's ad server. Hitting the "Back" and "Forward" buttons on the browser compounds the counting and rotation problem. Because Net Gravity does not get called when the page is in browser cache, it can not call AdForce (Net Gravity uses the Ad Insertion technique - See explanation above). Neither ad server will count these impressions and banners will get served from cache.			Cache-busting the AdForce tag fools the proxy server into thinking the AdForce tag is unique - therefore, everytime the AdForce tag is called, it not only counts each impression, it also rotates the banners as expected. So, as users navigate through the site, and as long as Net Gravity is called and given the chance to serve the AdForce tag, the banner will be pulled from the ad server and counted by both Net Gravity and by AdForce. However, because Net Gravity uses the Ad Insertion method, all impressions are pulled from cache when users hit the "Back" and "Forward" button on their browsers.			
Conclusion:		Sites numbers end up being higher than the third-party ad server's. Because agencies use the third-party server for report consolidation, they often insist that their numbers be "the" numbers reported to buyers. Sites, therefore, end up offering "make-goods", even though they have actually served all of the impressions. This scenario penalizes the website and favors the buyer.			The Counting discrepancy goes away because the third-party ad servers numbers come much closer the the site's numbers. Some cached impressions are still not counted so buyer gets the benefit.			

The Adauction.com Ad Serving Philosophy and Methodology

For the Media Buyer:
Adauction.com works with both AdForce and DART. Both are ad servers that use the cache-measured methodology. As discussed in detail in the table above, this methodology counts all impressions that are seen. We believe this is the proper way to count because it favors neither the buyer nor the publisher and it basically represents reality (or as close to reality as we can get)1.

There is some fallout from this decision however, of which you should be aware. When you buy through us, you may experience a lower click-through rate than what you are currently used to. This is especially true if you buy inventory at our auction that you once purchased directly. If that particular site was using, for example, Net Gravity (an ad decision maker that uses the ad insertion methodology), and the site was not busting cache on their side, your performance stats most likely did not include all of the impressions that you actually received. And unfortunately, your click-through rate was probably skewed on the high side.

We believe that the industry is moving toward a standard that states that every impression seen should be counted. Ad servers that use the cache-measuring technique count impressions in this manner. In our opinion, this methodology is fair to both buyers and publishers because buyers receive what they paid for and publishers don't have to over-deliver. This methodology also delivers statistics that are close to reality. So, advertisers get a realistic picture of their marketing campaign.

For the Publisher:
Our tag is usually served as a redirect from either a CGI program or an off-the-shelf, in-house ad server like Accipiter or NetGravity. This means that our delivery numbers are always lower than those of the publisher (for reasons, see Scenario 5 above as well as Diagram B) and because we use our statistics to report back to the buyer, we inevitably end-up asking that the site make-up for their "under-delivery".

In order to rectify this situation, we strongly encourage all of our publishers to implement cache-busting code on our tag. This ensures that all impressions counted by your ad server are also counted by our ad server.

Conclusions:

Certain counting methodologies - specifically static ads, ad insertions and dynamic ads - ignore cached impressions. The end result is that publishers end up over-delivering impressions in order to make-up for those that are never counted. These impressions don't show up on the buyer reports and so, in effect, buyers get more impressions than what they paid for and they receive performance results that are skewed and misleading.

Cache-busting helps but does not solve the entire problem because impressions viewed when a user hits the "Back" or "Forward" button on their browser are not always counted, even when cache-busting code is put in place.

Cache-measured ads, for the most part, count all of the impressions that are seen, even when the user hits the "Back" or "Forward" button on their browser. However, when the tag of a cache-measured ad server is served as a redirect from any other ad server, the publisher must still append cache-busting code (scenario 5). This is because the redirect gets cached, while the publisher's call does not. The publisher will therefore report higher numbers than the third-party ad server, resulting in counting discrepancies.

Final Statement:

Many within the on-line advertising industry acknowledge that a counting standard must be set before the industry can progress much further. In our minds, this is imperative. However, until that standard is established and widely adopted, we will continue to grapple with the issues of impression counting. Our hope is that by thoroughly explaining the complexities of counting and caching, we can set expectations appropriately and at the same time, work towards establishing the standard this industry so desperately needs.

Notes
1) Newer browsers and many proxy servers are getting smarter about cache-busting. After about 6 reloads, a "smart" browser will recognize that the call to the ad server is actually the same each time, even though each call is marked with a unique number. When a "smart" browser recognizes this circumstance, it will pull the banner from cache.

Acknowledgements:

This document could not have been written without the help and expertise from the following individuals:

David Kopp – AdForce
John Willis – AdForce
Jeremiah Budzik – DoubleClick DART
Matt Artz – Adauction.com
Charin Kidder – Adauction.com
Leesa Lee – Adauction.com

Banner advertising more effective than tv or radio in luring web shoppers, according to Andersen Consulting survey

Title: Banner advertising more effective than tv or radio in luring web shoppers, according to Andersen Consulting survey
Author: Andersen Consulting
Abstract: While most Web site banner ads are price-oriented, less than half of experienced web users cite price as the primary driving their web purchases, according to the survey.

Copyright: ©1996-1999 Andersen Consulting. All Rights Reserved.
Biography: Andersen Consulting is an $8.3 billion global management and technology consulting organization whose mission is to help its clients create their future. The organization works with clients from a wide range of industries to link their people, processes and technologies to their strategies. Andersen Consulting has more than 65,000 people in 48 countries. Its home page address is http://www.ac.com/

NEW YORK, November 24, 1999

Sending a serious signal to online businesses and countering conventional wisdom, 25% of experienced Internet users said that banner advertising drove them to shop online, beating out newspapers or magazines ads (14%), television commercials (11%), radio spots (4%), and billboards (4%), according to a nationwide survey of nearly 1,500 experienced Internet users by Andersen Consulting.

While most Web site banner ads are price-oriented, less than half of experienced web users cite price as the primary factor driving their web purchases, according to the survey. Instead, consumers rank convenience, time savings and site security as more meaningful factors in driving their online shopping.

"Promoting rock-bottom prices won't ensure online purchases or repeat business. Savvy net consumers consider the full shopping experience - from first click to home delivery," said Mary Tolan, Andersen Consulting's Managing Partner, Retail. "Consumers want to be able to return merchandise, and they don't like paying for delivery, which suggests a new set of marketing messages. eTailers can choose to promote themselves to the death over decreasing prices, or they can deliver holistic services that provide real value to consumers and generate real profits," said Tolan.

Experienced Internet Users Buying More Online This Holiday Season
According to the study, 82% of experienced web users are going online to buy gifts this holiday season up from 56% last year. In addition, 55% of holiday web shoppers say they will make at least one holiday purchase online this year, up from 37% last season. Over half (58%) of all online shoppers say they will spend less time shopping through traditional methods, such as malls and catalogs, in favor of the web this holiday season. Retailers seeking to establish their online brand during this crucial holiday season, must provide superior customer service and

Banner advertising more effective than tv or radio in luring web shoppers, according to Andersen Consulting survey

unique value to attract and retain customers, the study suggests. For example, only 34% of online holiday shoppers will know exactly what they want to purchase when they go online this season, while 63% will have a gift idea, but will not know exactly what they want to purchase. Online retailers offering advice and personalized gift suggestions will be providing a sorely-needed service for online shoppers, boosting customer loyalty and expanding bottom line profits, the study noted.

Respondents also said they plan to use online sites that have the following characteristics:

- 80% plan to use sites that offer good value
- 74% favor sites that provide ease of use and convenience
- 65% want to shop at sites that have merchandise in stock and provide quick delivery
- 64% will frequent sites that offer variety and an assortment of products

The survey also identified the factors which deter online purchases. Beyond the often mentioned concerns of privacy and security, Internet users who have not bought online rated the following issues extremely influential reasons for shunning eCommerce:

- 46% have a concern over returning unwanted goods
- 44% enjoy the traditional brick and mortar shopping experience
- 42% say web sites require too much information for purchases

- 41% do not want to pay for delivery

"eTailers are judged by their entire capability rather than just a cool web site or a slick ad campaign," Tolan said. "Service still matters, whether you are a 'click & mortar' or traditional retailer. The magic is what happens after a sale is recorded: having the right stock, the ability to ship in real time and being able to provide post-purchase support will determine online survival."

Study Methodology

A total of 1,472 Internet users were sampled. The surveys were completed over a three-day period and the data weighted to reflect the adult online population in the United States. The weighted sample, comprising 1,424 respondents, has a margin of error of ± 2 %. A post-holiday survey will be conducted among this same population to determine how anticipated behaviors and general online shopping expectations matched real experiences.

Chapter 5: Tips

9 Ways to Write Sure-Selling Ads

Title: 9 Ways to Write Sure-Selling Ads
Author: Binnie Perper
Abstract: You must communicate the genuine benefits and authentic value of your product or service to prospective buyers in a compelling and persuasive way.

Copyright: © 1998 by Writing By Design. All rights reserved.

What makes your ads, sales letters, web sites and other marketing materials as hard-hitting and sure-selling as possible?

Certainly, you must start with a product or service that delivers genuine benefits and authentic value to your customers and prospects. And each individual marketing piece must concentrate on one single big idea. It also helps to illustrate the benefits or product in a compelling way. Lastly, you must establish a program of regular communications with prospects and customers, because out-of-sight-out-of-mind is nowhere more true than in marketing.

But those are just the beginning-now you must COMMUNICATE the genuine benefits and authentic value of your product or service to prospective buyers in a compelling and persuasive way.

Follow these ad writing tips to do that:

1. WRITE AS YOU TALK.

Your company or product has a "personality," and that character has a voice. Use that voice in your copy. Make it sound like written speech, a printed conversation between your company's or store's "personality" and your buyer. Your tone, style and particular word choices will be different, for example, if your company belongs to the Mercedes crowd rather than the Kia crowd, or if you are a trend-setter versus a safe, traditional operator.

2. WRITE DOWN FEATURES AND BENEFITS BEFORE YOU START.

People don't buy what your product or service is - they buy what it does for them. Are your customers buying color printers, or the enhanced impact of color presentations? Are they buying a furnace, or the reliable, low-cost comfort it brings?

Are they buying "friendly service," or the education that service brings them, and the time and effort that service saves them? Writing down features and benefits forces you to concentrate on your product or service from both your point of view (features) and your customer's (benefits).

3. WRITE FROM THE "YOU" POINT OF VIEW.

Corporations don't sell. People do. Make every communication a one-to-one conversation, emphasizing how important the individual customer is and what you can do for that customer. Here are some phrases to keep you on track:
- You can get ...
- You'll find that ... it brings you....
- As you can see ...

4. VARY SENTENCE LENGTH.

Short sentences (2 - 10 words) move the story along quickly. But too many in a row make your message sound choppy. Long sentences carry your message along at a more leisurely pace, but can be confusing and cumbersome. By the way, it's okay to write one-word, two-word or incomplete sentences - "Yes, you do!" or "More?" or "Honestly!"

5. USE THE ACTIVE TENSE.

Results and actions don't happen by themselves. People act, and results occur. "Protect your family and valuables from burglars with an ON-GUARD alarm" is much more powerful than "Loss of valuables, and injury or loss of life, are averted with an ON-GUARD burglar alarm."

6. MOVE ON WITH THOUGHT CONNECTORS.

Keep your readers engaged by guiding them seamlessly from one idea to the next. Use natural transitions, like these:
- What's more ...
- In addition ...
- Better yet ...
- Equally important ...
- Remember ...
- Plus ...
- You can also ...
- More? ...
- First ...
- Lastly ...

7. STICK TO THE "RULE OF THREE."

A series of 3 adjectives, adverbs or ideas has more rhythm and balance than two or four. For example:
- The woods are lovely, dark, and deep.
- Broadcast fax is a fast, easy, and convenient way to deliver messages to customers, suppliers and the press.
- The updated program gives you a flexible merchandising kit, popular products and valuable guidance in setting up your product display center.

9 Ways to Write Sure-Selling Ads

8. SPELL OUT SPECIFIC ACTION.

Don't leave one action to chance - tell prospects and customers exactly what you want them to do ---- each and every step.
- Fill out the [form], put it in the postage-paid envelope and drop it into the nearest mailbox.
- Call toll-free 800-123-4567 today!
- Complete the order form, especially the mandatory items in red, then just click SEND at the bottom of the page.

9. RETURN TO THE OPENING THEME.

Conclude your ad, sales letter and brochure by bringing your message back to the theme or idea that you used to open it. This gives readers a sense of completion, and wraps the various elements of your sales message into a nice, neat, complete package.

The Ultimate Internet Advertising Guide

10 Tips to More Effective Banners

Title: 10 Tips to More Effective Banners
Author: Nick Bullimore
Abstract: Don't just sit around waiting for people to come to your site. You should be advertising it with banners. The banner should be animated. And yes, bigger is better and less is more.

Copyright: © 1999 Attard Communications, Inc.

Don't just sit around waiting for people to come to your site! You should be advertising it with banners? Use these Ten tips to create your own effective banner.

1. Create a banner immediately

If you do nothing else Utilize the most used phrase in cyberspace: Click Here. The key to effective advertising on the World Wide Web is getting potential customers to react, therefore the use of this strong clear cut statement which gives a call for action, together with an obvious immediate message, initiates the first and vital step clicking in your direction....

Equally words that tempt/entice the surfer into your world, such as Enter Here, Press Now! have shown that their use can increase response rates dramatically. Other good words to use are Free and This is (Your Last Chance) a feeling of urgency is created with the use of these words so make sure you can put credence to your claims. Flamers can take great offense at false claims and cause you all sorts of embarrassment so don't make these claims lightly prudence is required in their use...

2. Movement is the Key to Grab Attention.

ZD Net (a popular US Site) recently conducted a survey and found that animated ads generated 15% higher click thru rates than their counter parts, static banners, it was found that in some cases they were as much as 40% higher. Coupled with an intelligent but witty message and the appropriate design this form of advertising can work very well. As a seller of Ultra High speed Internet Access found out, combining a basic animation consisting of two scrolling bars filling up extremely slowly, together with a strong but simplistic message to make it go faster, 'Click Here'

This ad pulled an initial 19% response rate and was still averaging rates of 15% three and four months later. Most banner

10 Tips to More Effective Banners

averages 2% so their use of this method is a clear demonstration of the power of animation. Take Note be very careful when using banner animation as if it takes too long to download, you will only encourage the interested party to move on elsewhere, so make sure to check your banners loading time (on average a surfer will move on after 6 seconds if your page does not load efficiently).

3. Audience Involvement

Net Media is one on it's own unlike other offline media the customer has to come to you when looking on the web.. the best way to achieve this result is invite your target market either him or her to participate in some activity carried on at your site. perhaps a Game, Quiz, Give-away, work on the premise that people like to play games. Simple games such as noughts and crosses have been found to increase click-through rates. However Competitions have been found to offer virtually no incentives and really should be avoided.

4. Something for FREE

Give something away FREE the internet is unique in that the majority of its users have an active dislike at to big business prying into their world. Offer Net users something Free to get them to respond to your commercial pitch. Ezines (e-mail newsletters) and product updates can work very well with a targeted group, screen savers can produce good results as they can be used with subliminal advertising.

Free downloads can of beta software can and do create effective responses.

5. Don't Include Your Brand Name

AOL and recent Internet surveys by the Internet Advertising Bureau have found that Net users have a high level of recall of online ads coupled with increased brand awareness if the banner announces a new product or service, you should Not include the Brand Name. Viewers often assume they already know everything they need to know about the brand and totally ignore it. Please Pull them to You Don't Push them away... Quality information ...is the key..

6. A Must Change Your Banners Frequently

It is called Banner Burnout and it is rapid, after about only 200,000-350,000 impressions, the response rates can drop by half the second to the third time a punter sees the ad. Here is a Tip have at least three to four versions of a banner placed on various sites of importance. This enables you to note which of the four achieves the most response and where they are best positioned. Rotating banners and removing bad performers is tantamount to creating an effective and efficient banner campaign.

7. Target Your Market

Attach your banner to the most relevant key words, it will increase the chance of its location and your sites location in the

10 Tips to More Effective Banners

search engine databases. Search engines are where most surfers start to look when wanting information, if possible you should consider possibly sponsoring certain keywords or phrases with various search engines. For Example this would enable a lawn mower manufacturer's banner to appear whenever the word 'garden' is entered.

Two other choices offered by the search engines are category sponsoring, which speaks for itself and Run of Site advertising (ROS) which best suits products with a broad appeal or brand builders. Advantages with Run Of Site advertising is cost, this method can be as much as 80-85% cheaper than targeted ads.

8. Grabbing People's Attention

There is nothing sells better than Sex and Intrigue. Recent surveys confirm that Sex on the Internet is the most popular topic followed closely by conspiracy theories.

This is not so surprising as in general the human race is naturally concerned first and foremost with its own reproduction or elimination.

With regard banner advertising, a number of well chosen words are going to tempt even the most ardent Net-boffin, put together with some complimentary sounds results should be imminent. Whether you use just the click of a mouse button or a repetitive drum roll, attention is your aim.

9. Yes Bigger is Better and Less is More

Somewhat of a contradiction in terms but the results do work. Keep your banner size small with regard to bytes say under 10k-15k this will speed download, and then use wider banners - either 468 or 500 pixels wide.

It is a known fact that smaller banners do not generate as many responses as wide ones, many of the design functions work far better in a wide format. I have found that Yellow and Green are my preferred colors to use when creating banners.

10. Location, Location

Obviously at the Top of the List are search engines as they are the most visited sites on the Net as this is how surfers negotiate their initial Net enquiries but the beauty of the Web is the diversity of it's members and their interests. Really and Truly there are only three ways of getting your banner on to the best and most appropriate sites: 1. Sit down, spend a lot of time searching out the best sites that fit your target market, 2. Use and Ad Network (Paid or Free it is your choice) or 3. Employ an Ad Agency.

Link Exchange is a good start, join up for Free or Pay they place your banner on sites that are deemed appropriate. They operate a 2-1 system one of your banners is placed on appropriate sites in exchange for two other advertisers banners on yours).

You must remember that You Don't have much control over the last two methods.

10 Tips to More Effective Banners

For an instance Take double-click, although it has a high media profile it only represents about 75 Web sites among which are the search engines. Although it promises to deliver to a much sort after Targeted Audience, it depends on using ISP addresses as its main database. Taking this into consideration and how many users an ISP such as Demon represents you are grasping the problem with this kind of marketing.

Companies are looking toward sponsorship of specific web sites that carry relevant information and suitable material, this enables closer branding of content and product this being an alternative to the 'scattershot' technique. Although, editorial and subject targeting systems are rapidly being developed. The choice this will provide, marketing your banner to the exact person, who wants to see it just when they are ready to purchase your product is soon set to arrive.

Chapter 6: Webvertising In-Depth

Designing Catchy, Effective Banner Ads

Title: *Designing Catchy, Effective Banner Ads*
Author: *Meredith Little*
Abstract: *When you create your banner ads, keep your options open. Ads are no longer just colored boxes with some text. By following some guidelines, you can quickly create ads that stand out on the page and still load fast.*

As a Web designer, you face a number of challenges when creating Internet banner advertisements. You have a very small space-usually about 468 pixels wide by 60 pixels high-in which to convey your message. That message has to be legible, and it has to catch the reader's eyes on what is often a very busy page, all while keeping file size to a minimum. That's a lot to ask from such a small graphic!

Fortunately, you can follow some easy guidelines to create ads that are catchy, effective, and easy to make. In this article, we'll walk you through the following eight steps for creating your ads:

1. Determine the components.
2. Create a canvas you can work on.
3. Choose your colors wisely.
4. Arrange your elements without confining yourself to the box.
5. Make sure your text is legible.
6. Keep your file size down.
7. Make your ad jump off the page and catch the reader's eyes.
8. For measurable success, make sure your ad links to a targeted area.

You'll find that these tips will not only improve your banner ads but will also help you organize and create those ads quickly. Just follow through the first four steps for quick creation methods.

The trick is to make your ad different but still keep it simple by varying your layout options and experimenting with various effects. Figure 1 shows an example of a few of our guidelines in action: functional components, browser-safe colors, easy-to-read text, a layout that draws the reader's eyes into the ad, and boundaries formed by graphic elements instead of a static box.

Designing Catchy, Effective Banner Ads

Figure 1

Step 1:

Sometimes the most difficult part of designing any graphic is the creative aspect. If you're like most designers, more than once you've set up your blank Photoshop document and thought, "What now?" You can quickly bypass this version of writer's block by assembling the following staples of any banner ad:

- The main message-What's your ad selling? Be specific here; few readers will click on a cryptic ad, regardless of how cute the message is.
- Company logo-Who's doing the selling?
- Graphic-You need something that will attract the attention of your market.
- The URL-If the reader prints out the page or decides to visit the site later, you want to provide a way to reach the site.
- The words Click Here-Yes, it's obvious, but we promise that your ad will garner a higher click rate with these two magic words.

Now gather these elements in one place, as shown in Figure 2 (they can be in separate documents if you like). Think of this as your palette.

Having all your elements in one place should quickly eliminate your writer's block and spark the creative process. Although we won't discuss animated ads in this article, if you're assembling these components for an animation, make sure that as many elements as possible are in the last frame of the animation.

Another advantage to grouping all the components before you start laying out the ad is that it can help you avoid a busy ad. If you stick with these basic elements instead of just adding what you want as it occurs to you, your message will stay clear and your ad will be more successful. Or you can dispose of one or two items at this point if they make your ad too busy.

Designing Catchy, Effective Banner Ads

Figure 2

Step 2:

Once you've gathered all the elements, create a Photoshop document that's 468 pixels wide by 60 pixels high. Such a small document can be limiting, so it's a good idea to give yourself more room. But before you enlarge the canvas size, create a guide border of that 468 x 60-pixel space (or whatever dimensions your ad must be). Be sure to create the border in such a way that you can easily delete it from the finished image.

One way to do so is to duplicate the Background layer, and select all ([c]A on the Mac, or [Ctrl]A in Windows). Choose Select>Modify>Border. Set the Width to 1 pixel and click OK. Then press [option][delete] on the Mac ([Alt][Delete] in Windows) to create a 1-pixel-wide border at the edge of your image. As you work, don't place any of your elements on this layer. When you've finished laying out your ad, you can simply delete the layer.

Now enlarge the canvas size to give yourself enough room to work: From the Image menu, choose Canvas Size, add about 100 pixels to both the Width and Height, and click OK. At this point, you should have a canvas like the one shown in Figure 3, which provides you with plenty of room to arrange your elements and a guide for the final size.

Designing Catchy, Effective Banner Ads

Figure 3

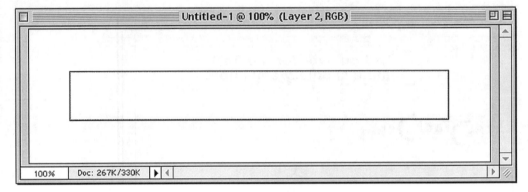

Step 3:

Once you've assembled your elements and prepared your canvas, you need to choose your colors. Your first consideration is the color of the background on which the ad will appear. If you don't have this information, it's best to stick with black or white.

When choosing colors for your ad, always make your selections from the 216-color browser-safe palette, which contains the 216 colors that display correctly on any platform. You can find this palette in the Goodies folder of the Adobe Photoshop CD. You can also download it from any of several online sites—just perform a keyword search with any search engine.

To correctly use the palette, you must get rid of the existing colors in the Swatches palette, not just add the new colors to them. To load the browser-safe palette correctly, display the Swatches palette by choosing Show Swatches from the Window menu, and select Replace Swatches from the dropdown menu, as shown in Figure 4. In the resulting dialog box, locate the new palette and click Open.

Figure 4

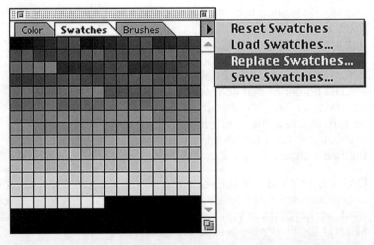

Designing Catchy, Effective Banner Ads

Figure 5

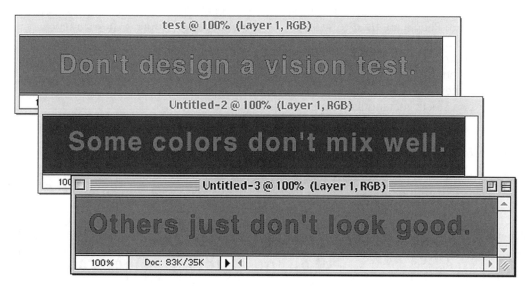

As you choose your colors, be aware of problematic color combinations. Figure 5 shows a few examples of color combinations you should avoid.

Step 4:

Now that you've chosen your colors, it's time to start arranging the elements on your canvas. If you're in a hurry, you can simply drag the elements over from where you gathered them and arrange them, as shown in Figure 6. Just get rid of your border layer and crop the image down to 468 x 60 pixels.

Figure 6

Designing Catchy, Effective Banner Ads

However, if you stop now, you're passing up one of your best chances to make your ad stand out. People are used to seeing ads as long boxes of color, so you should try not to confine your ad to the box. Although you do have to stay within the size limit, there's no rule that says you have to stick to a strictly rectangular shape.

Think of giving that box irregular borders by using elements of your design to define the boundaries. Figure 7 shows a few examples of how you can design "outside the box."

Figure 7

Designing Catchy, Effective Banner Ads

With these examples, we used most of the elements as we created them in our initial palette. Of course, you can reposition or break up text to give yourself additional options.

Step 5:

With all the competing elements in a banner ad, text can easily become lost. In addition, the small text that banner ad formats require can make text difficult to read. But, unless you have a really catchy image, readers use the text as the basis for deciding whether to click on your ad. So here are a few tips for keeping your text legible. First, try not to use text that's less than 9 points; if possible, keep it at 10 or 12. Second, use sans serif type-it generally retains its legibility on the Web better than serif fonts.

You can also employ a couple of tricks in Photoshop to increase legibility. When you create your type, increase the spacing between the letters. You can also duplicate your text layer, which increases the color saturation of the text, as shown in Figure 8, without increasing the color depth.

You might also try selecting your text and choosing Filter>Sharpen>Sharpen. This doesn't always make text more legible, but it's worth a try.

Step 6:

Now's a great time to weigh your image-to find out what size GIF file your image will make at this point. Knowing this will tell you what further enhancements you can afford to make. Of course, if you started with a minimal number of colors, you should be in good shape.

Save your file and choose Image>Mode>Indexed Color. Use the smallest bit depth that doesn't compromise the quality of the image, then save the file under a different name in GIF format and check its size. Generally, you should keep your banner ads to about 10-12 KB, or 12-16 KB if you're using animation.

Figure 8

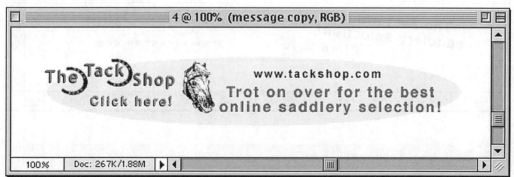

Designing Catchy, Effective Banner Ads

Step 7:

you have room for a few enhancements, think of ways to really draw the reader's eyes to your ad. Anything out of the ordinary will do the trick-creative borders, a dynamic layout, concentric circles around your focal graphic, or any of a number of text effects.

Figure 9 shows how you can lead the viewer's eyes into an ad by changing the perspective of the text (choose Layer>Transform>Perspective and experiment with the angle). The other Transform tools provide some interesting effects as well, most of which don't increase the color depth of your image.

Step 8:

Even the most catchy ad won't be successful if it isn't linked to the right place. Make sure that clicking on your ad will take readers to a page on which they can find exactly what your banner is advertising. For our example ad, a link to the home page of www.tackshop.com is just fine.

However, if our ad was specifically hawking a $600 cross-country saddle, the link should ideally jump to a page with information about that product. If the link must go to a different page, make sure it's easy for readers to find a path-preferably a short one-to the exact item being advertised. Otherwise, they'll quickly leave the site.

Conclusion

When you create your banner ads, keep your options open. Ads are no longer just colored boxes with some text. By following the guidelines we've shown you in this article, you can quickly create ads that stand out on the page and still load fast.

Figure 9

The Bigger Picture - Free Vs. Paid Advertising

Title: The Bigger Picture - Free Vs. Paid Advertising
Author: Internet Marketing Company
Abstract: There are really just two kinds of advertising – that which you pay for and that which you get for free. And both should have their place in your overall marketing strategy. The problem arises when one does not understand how and when to use each.

Copyright: © 1998-99 Internet Marketing Company. All rights reserved.

At the risk of sounding sensational, if you understand the concepts introduced in this article it could literally mean the difference between miserable failure and insane online profits for your business. I'm not talking about a magical formula that will guarantee success, but it's something that all online marketers need to understand.

Unfortunately, most do not - and therefore they spend every waking minute just trying to stay in business. And within a year, most small businesses give up any hope of making it big online - firmly believing that "this Internet thing just doesn't work."

Well I'm here to tell you that it does. In fact, the Internet is the best thing that has ever happened to the home-based or small business owner wanting to reach a global market. As with anything else in life, it's actually pretty easy to profit online if you know how.

When it comes to making lots of money, all that really matters is marketing. This is easily proven by the fact that you don't need a better product or a lower price to make more money than your competition. The truth is that you can literally make millions selling mediocre products - if you know how to market them effectively.

Of course, you should always strive to deliver quality products and great value to your customers - but the point is that marketing is all that really matters. It's what you need to be spending 90% or more of your time doing if you have any hopes of developing a profitable online business. Doing paperwork, building web sites, answering email, and processing orders doesn't help grow your business - only advertising does.

Advertising is the lifeblood of any business. If you do not learn how to advertise your products and services both efficiently and effectively, you won't be in business long. While the Internet has lessened or eliminated many of the costs normally associated with starting and running a small business, and it's now easier than ever, you'll never realize significant profits if

The Bigger Picture - Free Vs. Paid Advertising

you don't grow your business through effective marketing.

There are really just two kinds of advertising - that which you pay for and that which you get for free. And both should have their place in your overall marketing strategy. The problem arises when one does not understand how or when to use each.

Everywhere we look it seems that someone is claiming you can get rich online by taking advantage of the free advertising the Internet offers.

While it's a true statement, the misleading part is that it's almost always followed by something like, "We'll give you a list of over 984324984 places to advertise for free!" They normally go on to explain how anyone can get rich on the Internet because all of the advertising is free, and if you just send them $29.95 ... well you know how it goes.

Unfortunately, as anyone who has tried posting zillions of free ads on the Internet will tell you it just doesn't work that way. What you really need to know about posting free ads online is that for the most part it's a waste of time - unless you're targeting other marketers - because an overwhelming majority of people who visit free ad sites are other online marketers like you who are trying to promote their own business.

Effective "free advertising" strategies do exist, but they aren't what the average online marketer thinks of when hearing this often hyped-up phrase. In future issues we'll discuss things like networking, swapping links, joint-ventures, press releases, and the other "free advertising" strategies that you will want to pursue.

But back to the story at hand - free versus paid advertising.

Are you ready for the one concept that separates the men from the boys, so to speak, online? Here it is:

Paid advertising is the laziest way to promote your site!

You don't really think that Yahoo became the most popular web site on the Internet, getting millions of visitors per day, by using free advertising strategies do you? Heck no. Among other things, they spend millions of dollars per year on things like buying advertising on other big sites and advertising on both radio and television.

Now before you say you don't have millions per year to spend on advertising, allow me to show you how you can apply these concepts to your business on a smaller scale. The good news is that if you use the right approach you can start with a few dollars. It doesn't really matter what you're marketing, but let's assume that you sell widgets at your web site. How could you utilize paid advertising profitably?

Now before I scare you off with the thought of having to spend lots and lots of money on advertising, there's something that you need to realize. And it's a basic concept you should be able to immediately agree with:

The Bigger Picture - Free Vs. Paid Advertising

If you spend $50 on advertising and it generates more than $50 in net profits, that's a good investment.

Simple right? Well, through the use of proper testing and a "scientific" approach to advertising, there's no reason you can't turn your original $50 ad into millions of dollars in profits. It's just a matter of developing a system that creates a profit, and then reinvesting your initial profits back into your business in order to further expand your advertising.

Remember the widgets? Assume that through proper tracking of your web you're able to determine that 1 out of every 100 visitors to your site buys a widget. Let's also assume that on the sale of every widget you make a $50 profit. Do you see that any ad you buy which delivers more than 100 visitors to your site per $50 spent is a profitable investment?

Promoting a site should be approached with this type of a scientific or mathematical attitude. Any advertising you do is either profitable or not, and you need to know which it is so you don't waste time and money.

Here's a somewhat simplistic strategy that you could use to get started:

1. Use a combination of free or low-cost strategies to promote your web site, generating initial traffic which will serve as a starting point.

2. Through proper tracking of your site, determine the exact "value" of a visitor in terms of dollars and sense. This is absolutely critical.

3. Utilize paid advertising that proves to be profitable according to step 2, based on the value of a visitor and the number of visitors generated.

4. Through ongoing testing, tracking, and tweaking, try to increase the value of a web site visitor as well as response to your advertising.

5. Repeat steps 2-4 as necessary - or forever.

The power of this scientific or mathematical approach to web site promotion is that once you have completed the steps you will have an almost automated advertising campaign that can literally send you all the traffic you need. Rather than spending hours and hours each day promoting your site, you simply buy more advertising.

The bottom line is that there are only 24 hours in a day and there's only so much you can do during that time. Free advertising strategies can be effective, but normally, anything that's "free" is going to "cost" you time.

On the other hand, if you could spend $1,000 a day on advertising to make $1,000 a day in profits -- without spending hours doing it -- why wouldn't you just do that? It's not hard at all. Consider GoTo.com for example, where you can buy click-thrus for as little as 1-25 cents.

That fits the formula quite nicely. GoTo

The Bigger Picture - Free Vs. Paid Advertising

alone won't send you enough traffic to make you rich, but it's a good example of effectively promoting your site without spending much time doing it. There are lots of others.

Give this 5-step scientific approach a try. If you find you just can't seem to make the formula work, there can only be a few reasons for it - either you're not advertising in the right places, your web site isn't doing its job, your profit margins are too low, or your product or service itself is the problem. Figure out which it is and you can't fail.

Technical White Paper: Advertising on the Web

Title: *Technical White Paper: Advertising on the Web*
Author: *Tom Shields*
Abstract: *This white paper explores the technical support that exists for advertising on theweb, and compares several techniques that make advertising possible. It is intended for people who need to understand the technical limitations and advantages of the web medium. Conclusion: no one method is right for everyone. Many methods are extremely complex to implement and get right – make sure to budget an appropriate amount of time and development resources.*

Copyright: NetGravity, Inc.

Introduction

One of the first questions asked by any enterprise looking to take advantage of the growth of the world wide web is: how do we make money? For many information providers and entertainment companies, the answer to this is advertising. This white paper explores the technical support that exists for advertising on the web, and compares several techniques that make advertising possible today. It is intended for information providers, advertising executives, agencies, and rep firms who need to understand the current technical limitations and advantages of the web medium.

Overview

Advertising has taken many forms on the internet, but most of it occurs on the World Wide Web. Web pages provide a compelling medium for advertisers because they can be interactive - simply clicking on an advertisement can take the viewer to the advertiser's site, where they can get more information, or possibly order the product or service on line. Web pages also support graphics, formatted text, and other media designed to attract viewers to the advertiser's site. To take full advantage of advertising on the internet, it is necessary to understand some of how the web works.

Most of the ads seen on the internet now are "banner" ads, usually a simple image displayed across the top of the page. Other ad types, such as server-push images, Java, RealAudio, and Shockwave, are becoming more common. These ad types are only supported by a few browser types, albeit the most popular ones. The issues surrounding delivery of these ads are substantially the same as those for images, so many of the discussions below will apply.

How the IMG Tag Works

Browsers request data from servers, and the servers provide it. For the text part of

Technical White Paper: Advertising on the Web

each page, everything is sent at once. Other pieces of media, such as images, are sent separately, because they are often large. When browsers have image loading turned off, or if the image has been seen recently and is cached, they save bandwidth and do not even request them from the server. For browsers with images turned off, or without image support entirely, the text part of the page supports an "alternate text" label, that is displayed instead of the image. For example, a tag specifying an image with some alternate text might look like this:

<IMG SRC="hello.gif" ALT="Hello!">

Note that the actual image is not transmitted, just the name of it. The browser, if it has images turned on, will issue a new request to the server asking it to send the image (otherwise, it will just display the ALT= part). For a more complete explanation of how the IMG tag works, including Netscape extensions such as HEIGHT, WIDTH, and ALIGN, try here.

External Images

A lesser-known feature of the SRC= part of the IMG tag is that it does not need to refer to the same server as the page. Some sites get their ads from other sites by simply specifying an image on the remote site. For example, a search site might put up a Coca-Cola ad that looks like this:

<IMG SRC="http://www.coke.com/ads/current.gif">

This way, Coca-Cola could change the ad by putting a new image at "/ads/current.gif", and the search site wouldn't have to change anything. One non-obvious problem with this method is that it doesn't allow the alternate text attribute to be used. If, in the above example, Coke changes the "Coke Is It!" ad to "The Real Thing!", they would want to change the alternate text as well, for the benefit of those 10-20% of web users with images turned off. But, the alternate text is part of the document text, and would require changing at the source - the actual document at the search site. In many cases, this defeats the purpose of hosting the ad externally.

Dynamic Images

One of the interesting tricks that can be performed with the IMG tag is to specify a SRC= that is an executable script, instead of an image. This way, sites can implement a crude form of targeting. For example, an IMG tag might look like this:

<IMG SRC="/cgi-bin/select.pl">

This tag might use a script that could select the image based on the browser type or top-level domain. Here's a perl script that might implement this:

```
#!/usr/bin/perl

if ($ENV{"REMOTE_HOST"} =~ /.de$/i) {
    # the hostname ends in ".de" - Germany
    $filename = "german.gif";
} elsif ($ENV{"REMOTE_HOST"} =~ /.fr$/i) {
```

Technical White Paper: Advertising on the Web

```
    # the hostname ends in ".fr" - France
    $filename = "french.gif";
} else { # assume English
    $filename = "english.gif";
}

# read the entire file and write it to stdout
undef $/; # set mode to slurp whole file
open(FILE, $filename);
print <FILE>;
close FILE;
```

One major problem with this mechanism is the inability to select different media types, such as Java or Shockwave. The IMG tag is only capable of handling images, nothing else. Thus, this method cannot be used to display a Java ad if the browser is Java-enabled and an image otherwise. This mechanism also suffers from the alternate text problem described above.

Counting Ad Impressions

The most common method of counting ad impressions is to count the number of times the ad image is downloaded. Because of caching, browsers with images turned off, text-only browsers, and caching proxy servers, this number may bear no relation to the actual number of people who has seen the ad. Also, if the ads are not images - text, for example, or other media embedded in the document - this method does not work.

Another method is to count the number of times the page containing the ad was downloaded. This method counts cached views, but also counts browsers who have images turned off (and see the alternate text) the same as browsers who see the image. This method is more complicated to implement, because it is necessary to know what ad or ads are on a given page at a given time. When ads are rotated automatically, this information may be difficult or impossible to generate.

The actual difference between the two methods ranges from 20-40%, depending onthe site. Which method is more correct depends on who you talk to. Counting page views is more accurate, in that the number reflects the actual number of people who saw the ad in some form. However, the image counting method guarantees that the number of people who saw the image (not just the alternate text) is at least that number. Sites that charge on a per-impression basis wish to maximize their impression count, and therefore will use the page counting method if they can. Advertisers want to see high yield percentages (clicks divided by impressions), and therefore often ask for the lower impression numbers given by the image counting method.

Counting Ad Clicks

Counting ad clicks requires a bit of cleverness, because once the user clicks on a link, the browser goes straight to the link's location, and the original server never gets notified. The trick is to use a feature called "location redirect", which allows servers to notify browsers when a requested page has moved somewhere else. Sites make the ad link go back to the same server and run a script, which records the click and issues a location redirect to

Technical White Paper: Advertising on the Web

the advertiser's site. The browser automatically follows that, and the user never notices.

For example, a complete ad might look like this:

`<A HREF="/cgi-bin/redir.pl?http://www.coke.com/"><IMG SRC="/ads/coke.gif"></a>`

The "redir.pl" script might look something like this:

```
#!/usr/bin/perl

# open the log file
open(FILE,">>redir.log");

# get the URL from the query string
$url = $ENV{'QUERY_STRING'};

if ($url) {
    # I have a URL, so issue the location redirect
    print "Location: $url\n\n";
    print FILE "Got a click for $url\n";
} else {
    # no URL - error!
    print "Content-Type: text/html\n\nRedir error!\n";
    print FILE "ERROR: Got a click with no URL!\n";
}

# cleanup
close(FILE);
```

Note that if a site places the same ad in several pages, but wishes to see how many ad clicks occurred on each page, the site must encode some space information in the redirection string. The encoding might look something like this:

`<A HREF="/cgi-bin/redir.pl?FROM=Index.html&GOTO=http://www.coke.com/"><IMG SRC="/ads/coke.gif"></a>`

Static Advertising

Static advertising refers to ads that are placed directly into pages as they are stored in the filesystem or other repository. This method is often used in high-traffic sites, because no ad selection work is done when the user requests a page - it's all done beforehand, by placing the ads in the pages directly.

The advantage of this method is that the server load for advertising varies with the number of ads, and not with the site traffic volume. Thus, it is predictable and easy to manage. This method has the disadvantage that targeting can not be employed. Each ad is tied to a page, and that page will only serve that ad. Note that high ad rotation rates can still be used - many sites use 5 minute rotation periods - but the ads don't change for each request.

Static advertising is often implemented by placing a new tag or comment within the content pages to mark where the ad should go. Then a separate process is run to replace that tag with the appropriate ad. This process may be executed at regular intervals, to enable rotation of ads through the page. Of course, this process can get extremely complicated, for example when sites want to rotate hundreds of ads through thousands of pages, using proportional time periods to spread out impression counts. Getting the process right can be time-consuming and frustrating.

Static advertising is also hard to track. Sites with fast rotation periods have a difficult time employing the page counting method of ad tracking (see above), because they usually don't have good records of which ads were placed in which pages. Also, if they have staging servers, the time that the ad actually appears on the internet may be much later than the time the ad is placed in the page.

Dynamic Advertising

Dynamic advertising means performing some ad selection at the time when the user requests the page. This can take two forms: the ads can be placed in the page when the page is served to the user, or the img tag can refer to a script to choose the appropriate image (see above).

Dynamic advertising is desirable because the ad can be targeted to the user, instead of just the page. This is a powerful model, one that advertisers love because more targeting means more qualified leads for them, and sites love because they can charge more for it. The disadvantage, of course, is that it requires more computing resources, and may be extremely difficult to make fast enough for high-volume sites. The computational power required increases proportionally with the number of users of the site.

There are two main ways to perform dynamic advertising: as a server-side include, or during the page construction process. Sites that construct their pages dynamically, such as many search sites, may choose to embed the ad selection and insertion into their page creation mechanism. Some sites implement this by creating an ad server process, which is called from the page creation process, to create and schedule the ads. This mechanism allows separation of the functions of ad handling and page creation, enabling them to be able to upgraded independently.

Server-side includes are a mechanism implemented by many web servers that to allow additional information to be inserted into a web page before it is served to the user. Dynamic ads can be inserted into static pages by using a SSI to insert ads at the time the page is processed by the web server. The disadvantage of this method is that the server must process each page, looking for SSIs, before sending them to the user, thus slowing the server's performance. To learn more about SSI, see here.

Comparison of Methods

This section provides a brief summary of the above described options for advertising, and lists the pros and cons of each method. Also, a description of who might use each method is provided.

Technical White Paper: Advertising on the Web

External Images
Pros: No maintenance required by the site - put them in and forget them. An external agency can maintain and rotate the advertising, instead of the site.
Cons: The site has no control over what image appears. The site cannot provide appropriate alternate text for each ad. The site cannot track ad views or clicks. This method only works for image ads.
Who should use: Sites just getting started with advertising. Sites that don't need to control their advertising. Sites for which advertising is not their primary revenue stream.

Dynamic Images
Pros: Allows limited targeting. Maintains local control.
Cons: Overhead of running a cgi-bin script for every ad. The site cannot provide alternate text for each ad. The click URL redirector must be able to figure out which ad was displayed. This method only works for image ads.
Who should use: Sites with modest advertising needs. Sites that don't need to support other ad media types. Sites wishing to provide limited targeting capabilities.

Static Ads
Pros: No per-impression performance cost for advertising. Allows specification of alternate text for images. Allows other media types, such as text and Java. Maintains local control.
Cons: No targeting. Difficult to rotate. Requires maintenance of placement records for counting impressions.
Who should use: Sites with high volume. Sites that don't need targeting. Sites with the ability to build or buy technology to implement rotation and reporting.

Dynamic with server-side includes
Pros: Allows most kinds of targeting. Allows simple and sophisticated reporting. Can handle any media type. Maintains local control.
Cons: The most difficult to set up. Requires a separate ad server for reasonable performance. Requires server-side includes to be turned on, slowing server performance. Server load for advertising varies with the volume of users, not ads.
Who should use: Sites with static pages that need to do targeting. Sites with dynamic page constructors they have no control over. Sites that can afford the server performance cost.

Dynamic with custom insertion
Pros: Allows complete targeting. Allows simple and sophisticated reporting. Can handle any media type. Maintains local control.
Cons: Difficult to set up. Requires potentially extensive modifications to the page construction process. Server load for advertising varies with the volume of users, not ads. Not possible with static pages.
Who should use: Sites that are entirely dynamic. Sites with considerable programming expertise. Sites that can implement ad scheduling within their content generation system.

Conclusion

No one method is right for everyone. Many of these methods are extremely complex to implement and get right - make sure to

Technical White Paper: Advertising on the Web

budget an appropriate amount of time and development resources. There are many other issues involved in a complete advertising management solution that are left unexplored by this paper, including: inventory management, scheduling algorithms, validation, reporting, session tracking, user demographic capture, and sophisticated targeting. For more information about any of these topics. try NetGravity's website.

Generate and Tracking Response to Promotional E-mail

Title: Generate and Tracking Response to Promotional E-mail
Author: Michelle Feit
Abstract: With the low cost of e-campaigns compared to conventional direct mail, you can mail and test more. This article shows what's been working in e-mail campaign formats, copywriting, reply mechanisms, and response tracking.

Copyright: e-PostDirect, Inc.
Biography: Michelle Feit is President of e-PostDirect, Inc. an affiliate of Edith Roman Associates, a firm providing list brokerage, list management, and database and Internet marketing services.

A lot of marketers are looking these days to maximize response to Internet direct mail when upselling buyers, keeping in touch with customers, or generating new leads. This article shows what's been working lately in e-mail campaign formats, copywriting, reply mechanisms, and response tracking.

Picking a medium.

Most people think of e-mail as plain text messages, and the majority of the 3.4 trillion e-mail messages in 1998 were certainly sent in this format.

Straight text is the simplest, most popular, and easiest e-mail format. But it doesn't take advantage of the graphic and interactive capabilities of the Internet. That's why you might want to consider two other formats: enhanced graphics and visual mail.

With enhanced graphics e-mail, html is used to add color, fonts, different type sizes, drawings, photos, and other visuals to the e-mail message. You can send an e-mail that looks more like a designed Web or catalog page than a memo or letter. This is ideal for catalog marketers and others selling tangible goods. Visual mail (tm) goes a step further, adding motion and inter-activity to the e-mail. Graphics, animation, video, and audio are incorporated into a self-contained file that is sent as an attachment to an e-mail or a link to a Web page.

With inter-activity, you can create dynamic presentations that have much more impact than traditional static e-mail. For instance, if you are marketing a consulting service, your visual mail, when opened by the recipient, could show a short scene of two executives discussing the problem your service could help solve.

Prospects don't like to waste time downloading large files. Proprietary compression programs enable visual mail presentations to be "shrunk" to a manageable file size typically 200K or less.

Generate and Tracking Response to Promotional E-mail

Choosing a format and structuring the message.

Contrary to the popular misconception that people don't read anymore, real prospects with a genuine interest in your offer want product information and will read "long" copy even on the Internet!

Yet people are busier than ever, so your e-mail should use the following structure: At the beginning of the e-mail, put a "FROM" line and a "SUBJECT" line. The e-mail "FROM" line identifies you as the sender if you're e-mailing to your house file. If you're e-mailing to a rented list, the "FROM" line should identify the list owner as the sender. This shows the recipient that the e-mail is not spam, but rather a communication from someone with whom they already have an established relationship.

The "SUBJECT" line should be constructed like a short attention-grabbing, curiosity-arousing outer envelope teaser, compelling recipients to read further -- without being so blatantly promotional it turns them off.

In the first paragraph, state the offer and provide an immediate response mechanism, such as clicking on a link connected to a Web page. As in all response marketing, the easier you make it for prospects to respond the greater the responses rates.

After the first paragraph, present expanded copy that covers the features, benefits, proof, and other information the buyer needs to make a decision. This appeals to the prospect that needs more details than a short paragraph can provide.

Choose a style that fits your audience and your content. In some e-mail, this body of information is presented as a series of short paragraphs separated by lines or asterisks. Others present it as a continuous long letter.

Another successful format is the e-zine (Internet magazine), in which the e-mail resembles a short newsletter containing multiple items. Each short section covers one product or service. Each section has a link to a page on your Web site providing more information on that product.

Response options and measurement.

Every e-mail campaign should contain multiple response options for both Internet and non-Internet replies. Internet response options can include clicking on a link embedded in the text that sends the recipient to your Web site, or clicking on an e-mail addresses the recipient can use to send you an immediate reply. These can be your general addresses, but even better is a unique Web or e-mail address. This makes responses easier to track, and a unique URL has the added benefit of bringing the prospect to a Web page designed specifically for handling inquiries for the specific offer advertising in the e-mail campaign.

The most common non-Internet reply mechanism is a toll-free number to call. This can be your general toll-free number, but a special 800 number specifically for Internet response makes results easier to track. To maximize responses, include a special offer in your Internet e-mail, such

Generate and Tracking Response to Promotional E-mail

as a bonus gift, free shipping, or discount. Even better, make clear the offer is available only to customers and prospects receiving this special e-mail promotion.

With these reply mechanisms in place, you can get a good sense of the response your e-mail campaign is generating. One common measure is click-throughs the number of people who clicked on a URL embedded in the e-mail message.

Many click-throughs bring the prospect to a Web page presenting more information and encouraging further response, such as completing a registration form, downloading a file, or placing an order. Measuring how many click-throughs go to this higher level of response gives an even clearer sense of whether your e-campaign is working.

With the low cost of e-campaigns compared to conventional direct mail, you can mail and test more, and more easily. Do so. The options for designing and writing e-campaigns are becoming clearly defined, but what works best is not yet well known. Test offers, messages, formats, response mechanisms, lists, and list segments. Keep track of the results and you will gradually discover what your prospects and customers respond to in e-mail marketing giving you an edge in future success.

Why E-Mail Lists Have Come of Age

Title: Why E-Mail Lists Have Come of Age
Author: Michelle Feit
Abstract: You can get the same results from e-mail as postal mail, but quicker. The biggest hurdle in e-mail marketing is finding enough e-mail addresses to support great results on a large scale. Most companies only have a portion of their database coded with e-mail addresses and of those coded, as many as 30 percent returned as undeliverable.

Copyright: e-PostDirect, Inc.
Biography: Michelle Feit is President of e-PostDirect, Inc. an affiliate of Edith Roman Associates, a firm providing list brokerage, list management, and database and Internet marketing services.

Availability of e-lists is growing rapidly, from 300 six months ago to more than 1,000 today. The number of marketers using e-mail lists is growing just as rapidly. In the most recent quarterly survey sent out to direct mailers by Edith Roman Associates, more than 50 percent reported using e-mail marketing in their advertising mix.

E-mail marketing is getting positive results. Barry Green, vice president, circulation, for Hearst Publishing, has successfully integrated e-mail into his circulation efforts, as well as made his house opt-in e-list available for rent. Ivan O'Sullivan, president of Elron Software, claims, "I can get the same results from e-mail as postal mail, but quicker." Sue Butler, president of KIDASA Software, says, "We got a 10-percent response rate from Electronic Products [e-list] and no complaints," making it her company's top performing list.

The biggest hurdle in e-mail marketing today is finding enough e-mail addresses to support such great results on a large scale. Most companies only have a portion of their database coded with e-mail addresses and of those coded, as many as 30 pecent are returned as undeliverable. Keypunch errors, which account for many bad addresses, may be reduced by requiring double entry on e-mail addresses. To build and update their lists, many companies are taking measures to collect addresses with changes in printed forms, computer systems and operational procedures. These measures will give a return on investment as companies gain an easier, faster and less expensive way to communicate with customers and prospects.

With those measures in place, time is still a barrier. So what about all those customers and prospects on your house list without e-mail addresses? E-mail address data appending services are available, which can take a standard mailing address and add the individual's e-mail address, immediately increasing e-mail potential.

Why E-Mail Lists Have Come of Age

Once gathered, e-mail addresses can become a new source of profit as list owners create opt-in program and rent their e-list to approved partners and advertisers. While some point to the relative scarcity of e-mail lists currently available for rent, not very long ago there were only 300 postal lists available, and now there are more than 30,000. Direct mail grew into a multi-billion dollar industry by companies renting each other's mailing lists. List owners profited, list users profited, and this is where e-mail marketing is going today.

There is some hesitance about renting e-mail lists, but complaints have not increased and response rates prove these e-lists are working. List owners who have chosen to rent or share their lists are getting creative with their new profit source. Some are selling sponsorships on electronic newsletters; others are personally endorsing the advertiser's e-mail message. Most are controlling their own transmissions, but this will change as more e-lists come on the market and mailers' concerns for duplication increases.

Interestingly, the anonymity of the list owner in traditional postal mailings has not translated to the Internet. Recipients of the e-mail messages know who sent the message, why they were sent the message and, if they choose, how to be removed from the list. Promoting awareness of the list owner gives credibility to the message, which may be one reason these e-lists are getting such good results.

IT Training Center

Newtonstraat 37c 3902 HP Veenendaal Tel: +31-(0)-318-547000 Fax: +31-(0)-318-549000 Internet: www.scnedu.com E-mail: info@scnedu.com